Electrical Engineering Developments

nova
science publishers
www.novapublishers.com

Electrical Engineering Developments

Design and Simulation of Electrical Machines with MATLAB
L. Ashok Kumar, PhD, S. Albert Alexander, PhD,
Y. Uma Maheswari, PhD (Authors)
2021. ISBN: 978-1-68507-411-1 (Hardcover)
2021. ISBN: 978-1-68507-476-0 (eBook)

Planar Antenna: Design, Fabrication, Testing, and Application
Praveen Kumar Malik, PhD (Editor)
2021. ISBN: 978-1-53619-898-0 (Hardcover)
2021. ISBN: 978-1-68507-042-7 (eBook)

Philosophy of Power System Protection and Security: Computer Aided Design and Analysis
Muna Hamid Fayyadh, Samir Ibrahim Abood (Authors)
2021. ISBN: 978-1-53619-099-1 (Hardcover)
2021. ISBN: 978-1-53619-253-7 (eBook)

Phase-Locked Loops: Structure, Functions and Applications
Shambhu N Sharma, PhD (Editor)
2020. ISBN: 978-1-53618-338-2 (Hardcover)
2020. ISBN: 978-1-53618-490-7 (eBook)

Maximum Power Point Tracking: Background, Implementation and Classification
Maurice Hébert (Editor)
2020. ISBN: 978-1-53618-164-7 (Softcover)
2020. ISBN: 978-1-53618-279-8 (eBook)

More information about this series can be found at
https://novapublishers.com/product-category/series/electrical-engineering-developments/

Mykhaylo V. Zagirnyak,
Andrii P. Kalinov,
Anna V. Kostenko
and Viacheslav O. Melnykov

The Fault-Tolerant Control of Induction Motors

nova
science publishers
www.novapublishers.com

DOI: https://doi.org/10.52305/HBHU7375

Library of Congress Cataloging-in-Publication Data

Names: Zagirnyak, Mykhaylo V., author.
Title: The fault-tolerant control of induction motors / Mykhaylo V. Zagirnyak, Andrii P. Kalinov, Anna V. Kostenko, and Viacheslav O. Melnykov, editors.
Description: Hauppauge : Nova Science Publishers, [2022] | Series: Electrical engineering developments | Includes bibliographical references and index. |
Identifiers: LCCN 2022044933 (print) | LCCN 2022044934 (ebook) | ISBN 9798886972238 (hardcover) | ISBN 9798886973020 (adobe pdf)
Subjects: LCSH: Electric motors, Induction--Automatic control. | Fault tolerance (Engineering)
Classification: LCC TK2785 .Z3423 2022 (print) | LCC TK2785 (ebook) | DDC 621.46--dc23/eng/20221122
LC record available at https://lccn.loc.gov/2022044933
LC ebook record available at https://lccn.loc.gov/2022044934

Published by Nova Science Publishers, Inc. † New York

Contents

Preface

At present, variable-frequency, positional and tracking electric drives (ED), which perform controlled convesion of electric energy into the energy of mechanical motion of working bodies, are the main executive part of the automation systems of industrial mechanisms, machines and technological facilities. The scope of variable-frequency EDs is rather large: from high-power electric engineering to various areas of utilities and household devices. The use of variable-frequency ED makes it possible to reduce electricity consumption by 20–50% due to the application of mechanisms in which the motors are designed for maximum load, and the average daily load is 60–80%. At the same time, the operating conditions of motors and mechanisms in general are improved due to the exclusion of dynamic shocks, starting overloads and current limitations Thus, the use of variable-frequency ED allows creating an efficient energy-saving technology, the use of which enables not only saving electricity but also increasing the service life of the equipment. However, despite the simplicity and efficiency of modern variable-frequency ED systems, their proper functioning depends on the reliable operation of the motor, the three-phase off-line voltage invertor and the control system, as each of these components may malfunction. Electric motor malfunctions are the most significant share of the variable-frequency ED faults.

The most essential percentage of IM failures is caused by the damage in the stator power circuit. In most cases, such a damage develops gradually and eventually results in the complete failure of the electric machine. That is, IM operates for a long time with a damage in the early stages of its development. Such operating modes are characterized by the fact that the system does not lose its operability, but the indicators of the control quality significantly worsen. The energy efficiency of the energy conversion process deteriorates; the energy losses essentially increase and variable components of electromagnetic torque and active power consumption appear. Long operation of ED systems in such modes results in the further development of defects, and, ultimately, in the complete failure of the electric machine, which may cause accidents.

The termination of ED systems operation due to the above problems, on the one hand, and the ever-increasing trend of expanding the scope of variable-frequency ED with an IM in industry, on the other hand, turns the issue of their reliability and fault tolerance into the problem of paramount importance. The timely detection and elimination of IM damage in the early stages of their development can extend the technological equipment life and reduce the financial losses caused by unforeseen shutdown of the equipment due to technological failure or failure of IM. Thus, fault-tolerant control systems (FTC) are of particular interest. They are capable of detecting various types of damage at the initial stage and promptly adapt the control law in such a way as to preserve the ED operability for a long period of time until the possibility of IM repair or replacement occurs. That is, the most efficient use of the FTC system is industrial equipment, which should continue to operate despite the deterioration of dynamic performance and energy efficiency.

Thus, the task of developing methods to compensate for the impact of the IM stator windings damage on the dynamic and energy characteristics of VFED, which would be efficient and easy to implement, is topical.

Acknowledgments

We especially want to thank the TOV NVP ENERGO-PLYUS who made it possible to realize the experimental researchcontained in this book. It has given us the opportunity to solve the problem of compensating for the effect of damage to induction motor stator windings on electric drive dynamic and energy characteristics. Also, we appreciate the team of TOV NVP ENERGO-PLYUS who helped us to prepare the book manuscript.

Abbreviations List

ED	Electric Drive
IM	Induction Motor
FC	Frequency Converter
PWM	Pulse-Width Modulation
SVI	Self-Excited Voltage Inverter
EMF	Electromotive Force
FOC	Field-Oriented Control
DTC	Direct Torque Control
VFED	Variable-Frequency Electric Drive
MMF	Magnetomotive Force
FTC	Fault Tolerance Control
SC	System of Control
CFE	Complete Factorial Experiment
VCS	Vector Control System
AFC	Amplitude-Frequency Characteristic
PFC	Phase-Frequency Characteristic
ADC	Analog-Digital Converter
PC	Power Converter
MC	Measuring Complex
EC	Equivalent Circuit

Main Symbols

a	three-phase system operator
b	coefficient depending on the motor insulation class
C_{DC}	the capacitance of the capacitor in the DC link
di_s/dt	the derivative of the stator current module on the *i*-th measurement interval carried out on the leading time interval of the output voltage of the converter
e_A	electromotive force of the stator phase "A"
E_{max}	the maximum voltage of the DC link
E_{mean}	mean EMF values on the i-th measurement interval
E_{rr}	the calculated value of the rotor EMF for the rated mode of operation
f_{pwm}	PWM frequency
f_u	the frequency of the converter output voltage
h_i	the natural value of the variation interval
i	stator current vector module
i_0	the component of the zero sequence of the stator current
I_1	the amplitude of the stator current fundamental harmonic
$I_{A\ max,\ B\ max,\ C\ max}$	the maximum values of the stator phase currents amplitudes
$i_A(t)$, $i_B(t)$, $i_C(t)$	the instantaneous values of the stator current
i_A, i_B, i_C	stator phases current
i_a, i_b, i_c	rotor phases current
I_{cd}, I_{cq}	the compensation currents for torque-generating and flux-forming components of stator currents, respectively

(Continued)

$\widetilde{i}_d, \widetilde{i}_q$	the variable components of the flux-forming and torque-generating components of the stator current
I_d, I_q	the projections of the stator current vector on the axis of the rotating coordinate system d, q
I_{en}	the experimental values of current amplitude at different frequencies
I_{max}	the maximum value of the amplitude of the stator phase currents
I_{pn}	the calculated value of the current amplitude at the corresponding frequencies of the power supply network
I_{pwm}	the amplitudes of the current harmonics with PWM frequency
I_r	motor rated current
$I_{r\ mean}$	the mean value of the rotor current component
I_{r1}, I_{r2}	the active component of the stator current respectively before and after the jump
I_{rr}	the calculated values of the rotor current for the rated operating mode of the motor
$\vec{\mathbf{i}}_s$	stator current vector
$I_{s\ mean}$	the average value of the stator current
$I_{sd(ref)}$, $I_{sq(ref)}$	the initial reference signals of the flux-forming and torque-generating components of the current, respectively
$I_{sq(ref)}$	the initial reference signal of IM stator current along axis q
$i_{ss}(t)$	the instantaneous value of current via a semiconductor switch
$\mathbf{i_v}$, $\mathbf{i_z}$	the vectors of forward and reverse stator current sequences
i_{va}, i_{vb}, i_{vc}	the components of the forward sequence of the stator current
I_x	the reactive component of the stator current
i_{za}, i_{zb}, i_{zc}	stator current reverse sequence components

$I_{\mu\ mean}$	the mean values of the magnetizing current at the i-th interval of measurements
$I_{\mu r}$	the calculated values of the magnetizing current for the rated mode of motor operation
J	rotor inertia moment
J_A	objective function by amplitude quality criterion
J_φ	objective function by phase quality criterion
k	higher harmonics coefficient
k_c	skew coefficient
k_c^s, T_c^s	the transmission coefficient and time constant of the current sensor
kE_r	the module of the stator driven generalized vector of the motor rotor EMF
k_{fc}	FC transmission coefficient
k_{flp}, k_{fli}	the coefficients of proportional and integral components of the regulator of the rotor flux linkage
k_{fl}^s, T_{fl}^s	the transmission coefficient and the time constant of the flux linkage sensor
k_i	reduction coefficient ($k_i = I_n / T_e$)
k_m	modulation coefficient
k_{pc}, k_{ic}	the coefficients of proportional and integral components of the stator current regulator
k_r	scattering coefficient
k_{sp}, k_{si}	the coefficients of proportional and integral components of the regulator of the angular rotation frequency
k_s^s, T_s^s	transmission coefficient and time constant of the angular rotation frequency sensor
k_{w1}	stator phase winding coefficient
k_{w2}	rotor phase winding coefficient
k_δ	Carter's coefficient
L_A, L_B, L_C	the inductances of the motor stator phases
L_a, L_b, L_c	the inductances of the motor rotor phases
L_{DC}	inductance in the DC link
L_m	magnetization circuit inductance

(Continued)

L_s, L_r, L_μ	the inductance of the stator, rotor phases and mutual inductance
l_δ	the calculated length of the magnetic circuit
m	the number of output voltage pulses for a period
M	mutual inductance maximum value
m_1	stator phases number
M_{xy}	the mutual inductance of the motor windings x and y
$\widetilde{p}$	the variable component of three-phase active power consumed by IM
p(t), $p_{fc}(t)$	active power consumed from the network and from the frequency converter
$p_1(t)$	instantaneous active consumed power
$p_2(t)$	power on IM shaft
p_n	the number of IM poles pairs
$\vec{q}$	instantaneous reactive power vector
$q_1(t)$	instantaneous reactive consumed power
Q_a, Q_b, Q_c	spatial winding functions of phase windings
q_{in}	pulsing coefficient at the input of the filter
q_{out}	pulsing coefficient at the output of the filter
R_A, R_B, R_C	stator phases resistances
R_a, R_b, R_c	rotor phases resistances
R_e	motor total resistance
R_m	magnetization circuit resistance
R_s, R_r	the resistances of the stator and rotor phases
$R_{\Sigma n}$, $X_{\Sigma n}$	the experimental total values of resistances and inductive reactances of the windings of the researched motor at different reduced frequencies
s	pulsation leveling coefficient
s_n	conductor cross section
s_r	rated motor slip
T	the period of the variable power component signal
T_0	the conditional service life of the insulation at temperature $\Theta = 0$
T_e	IM electromagnetic torque

$T_e(t)$	the instantaneous value of the motor electromagnetic torque
$T_{ea\alpha}$	the constant component of IM electromagnetic torque
$T_{ea\alpha}$, $T_{eb\alpha}$, $T_{ec\alpha}$, $T_{ea\beta}$, $T_{eb\beta}$, $T_{ec\beta}$	the orthogonal components of the electromagnetic torque in IM three phases
$\vec{\mathbf{T}}_{eq}$	the vector of the instantaneous value of the electromagnetic torque reactive component
$T_{e\alpha}$, $T_{e\beta}$	the orthogonal components of the IM total electromagnetic torque
$T_{e\alpha 2}$, $T_{e\beta 2}$	the orthogonal components of the second harmonic of the IM electromagnetic torque
t_f	magnetization time with current limitation
T_l	load moment
T_{pwm}	PWM frequency period
T_r	rotor wheel time constant
T_s	the equivalent time constant of the induction motor stator
u	the module of the IM stator voltage vector
U_α, U_β, I_α, $I_{\beta cv}$	the projections of the voltage and current vector in a fixed coordinate system
$\dot{U}_{AB}$, $\dot{U}_{BC}$, $\dot{U}_{CA}$	linear voltages in a complex form
$u_0(t)$	the voltage in a neutral wire
$u_A(t)$, $u_B(t)$, $u_C(t)$	the instantaneous values of the stator voltage
u_A, u_B, u_C	stator phases voltage
U_d, U_q	projections on the stator voltage vectors on axes d and q
u_i	the coded value of the factor
U_{kd}, U_{kq}	compensation voltage
U_{kd}, U_{kq}	compensation voltage
u_m	the maximum allowable amplitude of the control signal
U_s	the constant component of the stator voltage
u_s	stator voltage module
$\vec{\mathbf{u}}_s$	stator voltage vector
$U_{t\,\max}$	the maximum value of the control voltage

(Continued)

w_1	the number of turns in the stator phase
w_2	the number of turns in the rotor phase
w_d	the number of turns in the damaged phase
w_h	the number of turns in the undamaged phase
x_{0i}	the value of the factor in the center of the plan
X_1	the coefficient of the stator winding data
x_i	the natural value of the factor
Y_A, Y_B, Y_C	phase conductivities
Y_e	load equivalent conductivity
Y_i	the instantaneous value of the real signal
$\hat{y}_i$	the value of the response function predicted by the regression equation for the *i*-th experiment
$Y_r(i\Delta t)$	the function that describes the instantaneous value of the calculated signal
Y_s	the average value of the real signal
Z_A, Z_B, Z_C	phase complex resistances
Z_e	the equivalent complex resistance of the load
γ	rotor rotation angle
δ	gap width
ΔA	static error in the signal amplitude
ΔI_p	stator current pulsations
$\Delta P(\varepsilon_w)$	losses in ED elements for the current degree of asymmetry
$\Delta p_{ad}(t)$	additional losses
$\Delta p_c(t)$	steel losses
ΔP_{Cu1}	stator copper losses in each phase of the rotor
$\Delta p_{Cu1\_id}(t)$	electrical losses in the stator copper in idle mode
$\Delta p_{Cu1\_l}(t)$	electrical losses in the stator copper in the current mode
$\Delta p_{fw}(t)$	losses in mechanical components
ΔPgr_i	heating losses after the *i*-th overhaul
ΔPgr_n	rated heating losses
ΔP_r	losses in the corresponding elements of the ED with symmetric IM
ΔP_{ss}	losses in FC semiconductor elements

ΔP_{Σ}	the total heating losses of the motor in the rated mode
ΔT	the interval between current samples
$\Delta\varphi$	static error in the signal phase
ε_w	the coefficient of the asymmetry of the stator phases
ε_{Ψ}	the correction coefficient of the flux linkage of the IM asymmetric phase
θ	the angle between axis α of the fixed coordinate system and axis d of the rotating coordinate system
$\Theta(i)$	the temperature of windings insulation after the *i*-th overhaul
$\Theta(\varepsilon_w)$	the temperature of windings insulation at the given degree of asymmetry
Θ_n	the rated temperature of windings insulation, determined by the IM insulation class
Θ_r	the rated temperature of windings insulation, determined by the IM insulation class
μ_0	magnetic permeability
ρ	resistivity
σ	scattering coefficient
σ_{bi}	regression coefficient error
τ	insulation service life
τ_1	the service life of the damaged motor insulation
τ_2	the service life of the damaged motor insulation when using the compensation system
φ_a, φ_b, φ_c	the parameters of the spatial arrangement of the magnetic axes of the phase windings in the cross-sectional plane of the machine
φ_{en}	the experimental value of the phase of the researched signals
$\varphi_{iA,\ iB,\ iC}$	the angles of the phase current vectors
φ_{pn}	the calculated value of the phase of the researched signals at different frequencies
$\varphi_{\psi A,\ \psi B,\ \psi C}$	the angles between the flux linkage vectors and the vectors of the torque-generating components of the currents
χ_A, χ_B, χ_C	phase electromagnetic parameters

(Continued)

$\chi_{AB}, \chi_{BC}, \chi_{CA}$	interphase electromagnetic parameters
$\Psi_{A\max, B\max, C\max}$	the maximum values of the amplitudes of phases flux linkages
Ψ_A, Ψ_B, Ψ_C	stator phase flux linkage
Ψ_a, Ψ_b, Ψ_c	rotor phase flux linkage
$\Psi_{\max}$	the maximum values of the amplitudes of phases flux linkages
Ψ_r	the module of the rotor flux linkage vector
Ψ_r	the value of the rotor flux linkage at the time of transition to the mode of maintaining the specified value of the flux linkage
ω	supply voltage frequency
Ω_2	slip frequency
ω_e	the electrical rotation frequency of the rotor
$\hat{\omega}_e$	the current value of the voltage frequency on the motor stator
ω_m	the output frequency of FC voltage
ω_s	rotor rated rotation frequency
ω_Ψ	the rotation frequency of the rotor flux linkage vector

Chapter 1

The Analysis of Induction Motor Damages and the Methods for Their Compensation by Means of a Variable-Frequency Electric Drive

1.1. The Trends of Development of General Industrial AC Variable-Frequency Electric Drives

At present, variable-frequency, positional and tracking electric drives (ED), which perform controlled convesion of electrical energy into the energy of mechanical motion of working bodies, are the main executive part of the automation systems of industrial mechanisms, machines and technological facilities. The scope of variable-frequency EDs is rather large: from high-power electric engineering to various areas of utilities and household devices. The use of variable-frequency ED makes it possible to reduce electricity consumption by 20–50% due to the application of mechanisms in which the motors are designed for maximum load, and the average daily load is 60–80% [1]. At the same time, the operating conditions of motors and mechanisms in general are improved due to the exclusion of dynamic shocks, starting overloads and current limitations. Thus, the use of variable-frequency ED allows creating an efficient energy-saving technology, the application of which enables not only saving electricity but also increasing the the service life of the equipment.

Combining the functions of technological process automatic control and of transformation of various types of energy in one technical device makes it possible to divide all ED systems into two basic classes according to the design principles and internal optimization [2]:

- "Power Eds," the main function of which consists in the controlled conversion of electrical energy into useful mechanical work. Such systems are designed for the most efficient use of the set capacity of ED power equipment (semiconductor power converters and electric machines) in steady modes with not very stringent requirements for the quality indicators of transient processes and control accuracy.

General industrial EDs and, in particular, the so-called energy-saving EDs operating mainly in static modes with a load that changes slowly can be identified among the "power" systems. The laws of control of such EDs are optimized by one of the common technical and energy criteria: the criterion of minimum current consumption, minimum power losses, etc.

- "Informational Eds," whose main function is extremely accurate and, consequently, high-speed reproduction of the coordinates of the mechanical motion of the working bodies of set spatial trajectories. Such systems operate mainly in transient processes, hence, the criteria for optimizing information EDs are obvious: maximum speed, static and dynamic accuracy.

In mass systems of variable-frequency EDs, currently used in industry, heating and water supply systems and operating mainly in long static modes with a constant load moment, the AC systems based on induction motors (IM) with short-circuited rotor consuming more than half of all produced electricity, are most common [3]. The widespread use of IM in mass ED can be explained by its relative cheapness and high reliability, simplicity of design, small size and moment of inertia of the rotor, etc. [4].

The emergence of a new element base and the use of modern microprocessor technology allows developers to create compact, multifunctional and highly efficient control systems for AC EDs, which meet the requirements of most technological processes [5, 6]. The advantages of such control systems include:

- high-quality control of the rotor rotation frequency;
- high torque at low rotation frequencies;
- low costs and high efficiency;
- high dynamic characteristics;
- steady operation with high-power motors.

A modern mass variable-frequency ED, as a rule, contains a frequency converter (FC) with a DC link, at the input of which there is an uncontrolled rectifier. The rectifier is loaded on a transistor self-excited voltage inverter (SVI) operating in pulse-width modulation (PWM) mode. This FC structure makes it possible, regardless of the mode of ED operation, to provide a fairly high power factor and high dynamic and static performance of the control system. With the creation of bipolar transistors with isolated gate (IGBT modules) and integrated circuits for their control (drivers), the scope of ED with transistor FC has become practically unlimited. Currently, transistor EDs with power from tens of watts to several MW have become common, and two main pathways based on the scalar (or amplitude) and vector principles of frequency control have been identified to control induction variable-frequency EDs.

In 1925, M.P. Kostenko was the first to solve the problem of maintaining the indicators of IM functioning with variable frequency, which were close to the rated ones. Since then many researchers have repeatedly addressed and continue to address the problem of energy optimization of ED static modes.

The simplest proportional law of the scalar control of the stator voltage amplitude, as a function of its frequency of the form $U/f = const$, is the most widespread in the practice of the design of the systems of automatic control of induction ED providing the set static indicators. However, with such a control law, it is impossible to simultaneously ensure satisfactory mechanical and energy characteristics of the ED in a wide range of changes in the rotation frequency and load due to the influence of the resistance and scattering inductance of the IM stator. In this regard, in the 60's it was proposed to use frequency-current control. In this case the stator phase windings form a three-phase system of sinusoidal currents, the amplitude, frequency and phase of which depend on the required values of the motor torque and flux linkage, and the current value of the rotation frequency or rotor position.

The choice of the control method and principle is determined by a set of static, dynamic and energy requirements for induction ED. If there are no special requirements for ED dynamics, and the static characteristics meet the conditions of the task, the use of frequency control in an open system is the simplest and most effective solution [7]. In such systems, IM frequency and supply voltage are formed in proportion to the control voltage in FC based on SVI. In its turn, the formation of the required static and dynamic properties of the induction variable-frequency ED is possible only in a closed control system [8]. Such coordinates may include stator current and EMF, IM main magnetic flux, rotation frequency and rotor frequency or the absolute slip. The choice of the backfeed signal is determined by many conditions: the nature of the load, the technical requirements for ED, the ability to use signals generated in other control circuits.

A significant disadvantage of the scalar control consists in the difficulty of implementing the desired laws of regulation of IM speed and torque in dynamic modes, which is due to incomplete taking into account the internal electromagnetic processes occurring in the IM. This, in turn, results in the deterioration of ED dynamic properties (reduced speed, increased overcontrol and oscillations of electromechanical processes) in transient modes of its operation, and in some cases can cause the loss of ED stability even in stationary modes. For these reasons, the scalar control principle is usually used in non-dynamic general industrial EDs (pumps, fans), characterized by operation in stationary operating modes or modes close to them, smooth start and braking, the absence of abrupt change in load.

The principle of induction ED vector control, proposed in 1971 in the papers of F. Blaschke, is the most promising at present. This control principle makes it possible to consider the IM as a two-channel object (analog of a DC motor with independent excitation) in the coordinate system oriented on one of the flux linkage vectors, and independently affect the longitudinal (magnetizing) and transverse (torque) components of the stator current vector for the control of the magnetic state of the machine and the electromagnetic moment, respectively.

When creating vector control systems for induction EDs, two fundamentally different approaches are used, which are called direct and indirect orientation of the vector of control influences in the direction of the motor magnetic field.

Direct Field-Oriented Control (FOC) [9] evaluates the components of the rotor flux linkage vector in a fixed coordinate system (α, β) based on the results of processing current information on variables available for direct measurements (voltage, current, and sometimes motor speed). They are used to determine the angle between axis α of the fixed coordinate system and axis d of the rotating coordinate system used in coordinate transducers. Systems with direct field orientation include only those in which direct flux measurement is performed with the help of certain sensors, or in which the flux is calculated according to the motor model, as this allows, as in direct flux measurement, creating a closed regulation circuit.

Indirect FOC [10] is performed without processing the information about the IM instantaneous currents and voltages by assessing the phase of the rotor linkage vector by integrating the sum of the electric rotation frequency and assessing the sliding frequency or adding the electric rotation angle of the rotor with the integral of the sliding frequency. The systems with indirect measurement include only systems in which the flux is not measured and calculated, but is formed by setting other variables.

The implementation of the principle of vector control became the initial prerequisite and basis for further use of other more complex control principles:

field-oriented induction machine control of its electromagnetic parameters of the mode or direct torque control of induction machine.

Despite the fact that technically (algorithmically) the principle of ED vector control is much more complex than the scalar control, the general global trend of modern variable-frequency induction ED is based on expanding the scope of the vector control. This is explained by the fact that due to complete control of the current value of the generalized vector of fundamental harmonics of voltage (current) of the machine stator, it became possible to adjust the instantaneous (current) values of flux linkage and electromagnetic torque of the induction motor (which is not possible with the scalar control). In turn, the regulation of the current values of flux linkage and electromagnetic torque of the IM is required when creating high-quality (with increased speed and accuracy) systems of variable-frequency EDs.

If the quality of variable-frequency ED with the scalar control is compared in stationary modes (provided the conditions of steady work of ED in necessary operation ranges of the change of the machine speed and loading are created) with vector control, the quality of regulation under the same laws of frequency control of the machine (characterized by static mechanical characteristics, the accuracy of maintaining the adjustable parameters of the ED mode, energy performance of the IM) for both principles of frequency control is almost equivalent. However, the main significant difference between variable-frequency EDs with the scalar and vector control is observed in dynamic modes. In this case EDs with the vector control (as opposed to the scalar control) provide significantly better dynamic properties (in terms of speed and ability to form the desired quality of electromagnetic and mechanical processes), as well as guarantee steady operation in wide ranges of rotor rotation frequency control and load torque.

The main and most rational area of application of the vector control principle follows from the above. It consists in its use for the variable-frequency induction EDs that are highly dynamic (operating in the conditions of sharp changes of the electromagnetic moment of the motor) and wide-range (with the expanded ranges of change of speed and the moment of loading). Increased requirements for their speed or for a part of the set (desirable) quality of the formed transitional electromechanical processes of ED are put forward.

The inaccuracy of the determination of the reference vector (rotor flux linkage vector) in the stator coordinate system is the main source of errors of vector control systems. Thus, the absence of rotor flux linkage sensors is compensated for by the calculation of differential equations, which include inaccurate and variable parameters of the IM, such as rotor and stator support, reduced moment of inertia of the rotor, the moment of loading. In this regard, vector control systems are very sensitive to the uncertainty of these parameters. Therefore, a number of methods have been developed that ensure the robustness (insensitivity) of the control system to the scatter of the parameters of the control object, such as: adaptive observers MRAS (Model reference adaptive system) [11], genetic algorithms [12], neural networks [13], fuzzy logic [14], variable-structure observers [15].

In order to avoid the shortcomings of vector control systems, such as the large amount of calculations for direct conversion from stationary to rotating coordinate system and inverse transformations and the delays in the formation of electromagnetic torque, a Direct Torque Control (DTC) algorithm was proposed in 1985 [16]. DTC algorithms are characterized by simplicity (absence of coordinate converters and no need to adjust the current circuits), robustness as to uncertain parameters, high speed.

DTC systems have a number of special characteristics: the presence of relay hysteresis regulators of the magnetic flux of the stator and IM electromagnetic moment in the system; the presence of a motor electronic model in the system for calculating coordinates; the presence of a tabular (matrix) calculator of the optimal vector of the motor voltage; the presence of the identifier of the phase sector, in which the vector of the motor stator flux linkage is at the current time; the explicit absence of the motor stator current regulators; no software PWM of the FC output voltage.

The presence of relay regulators in DTC systems is characterized by control influences rapid working off. However, the moment chaotic switching of the power switches in the inverter is possible in the direct control system in the steady-state mode. In order to reduce these drawbacks, one uses a high frequency switching of the power switches of the inverter by setting the width of the hysteresis loops in the flux linkage and torque regulators.

A significant disadvantage of the DTC system consists in the fact that the satisfactory quality of the transients is provided only if the error in assessing the stator resistance does not exceed 5%. Other drawbacks of such systems include the presence of pulsations in the electromagnetic torque and flux linkage, which reduces the accuracy of regulation, increases power consumption and increases the acoustic noise of the IM.

The use of sensors of position or angular rotation frequency of the rotor in the considered systems makes it possible to organize high-quality and relatively simple algorithmic control of IM. However, the presence of these sensors significantly impairs the performance of ED. Due to this, “sensorless” systems of variable-frequency induction ED are widely used in industry. The rotation frequency in them is assessed either on the basis of the information about motor current and voltage at the FC output [17] or, in general, using only current sensors [18], or the difference between the supply voltage frequency and the slip frequency assessment [19]. The main requirement for modern general-purpose EDs is to provide a relative static error of less than 5–20% in the speed control range: in the sensorless version – up to 100:1, in the presence of a sensor – up to 10.000:1.

Thus, the current trend in the development of variable-frequency ED is characterized by a large number of various control algorithms for IM, which have different qualities, and the choice of a control algorithm should be based on the requirements for the ED system. However, all these methods are more or less sensitive to the variance of the parameters of the IM electromagnetic circuit, which may vary depending on external influences, state variables, imperfections in manufacturing or repair technology, manufacturing defects, assembly or defects due to damage to the stator, rotor windings, mechanical components during long-term operation. These factors can result in changes in IM electromagnetic parameters both in all phases and separately in each phase, and are characterized by increased losses in the motor and the appearance of variable components of electromagnetic torque and power consumption, which significantly affects the ED performance.

1.2. The Analysis of the Causes for the Deterioration of the Operation Efficiency of the Variable-Frequency Electric Drive

The first reason for the deterioration of the efficiency of the VFED is the choice of the irrational mode of ED operation, for example, the operation of the motor in an underload mode, etc. Currently, there is a large number of papers and practical implementations of various methods for optimizing the ED energy consumption. Such methods include: the correction of the law of control [20-22]; the implementation of extreme control systems (by the minimum total motor losses, maximum efficiency or minimum stator current) [20, 23-24] and optimal control systems (e.g control by the flux) [20, 26].

One of the most important factors influencing the deterioration of the efficiency of AC ED is the asymmetry of the parameters of the induction motor [27, 28]. When implementing energy-saving technologies, IM is usually considered serviceable, electrically and magnetically symmetric. At the same time, statistics and industrial experience show that the fleet of electromechanical equipment of domestic enterprises mainly consists of electric machines that have been in operation for many years and have been repeatedly repaired. In such motors there is a constant process of aging and wear of the elements which is observed during all IM operation and causes a decrease in its operational reliability [39-31]. As a result of these factors, most industrial IMs have a certain degree of acquired asymmetry.

The asymmetry may be caused by such electrical damage to the stator windings as interturn circuits or breaks in parallel branches and elementary conductors of phase windings. Most often, such damage is the result of insulation failures and the occurrence of turn-to-turn short circuits. These short circuits entail local overheating, which can result in phase failure, parallel section or phase or interphase short circuit. The degree of asymmetry caused by such a damage depends on the number of parallel branches in the phase and the elementary conductors in the slot. It is known [32, 33], that in high-power IMs with a large cross-section of the phase windings conductors, to reduce additional losses from eddy currents, these windings are divided into a number of elementary conductors and coils are connected so as to form parallel branches. When one of the parallel branches breaks, the degree of asymmetry from such damage depends on the total number of parallel branches in the phase and the elementary conductors in the slot.

The analysis of the design schemes of induction motor windings showed that low-power motors ($\leq$ 15-20 kW) usually have a small number of parallel branches (2, 3), and a failure of one of them will cause significant asymmetry, which can be quickly diagnosed. In high-power ($\geq$ 100 kW), multipole motors ($2p \geq 4$), the windings are usually made with a significant number of parallel branches and additional division of the effective conductor into several elementary ones, connected in parallel [34-36]. In high-power IM the windings are most often made of rectangular wire with the possible number of elementary conductors in the effective conductor 2, 4, 6. Because of this, if one conductor or a parallel branch is damaged, such motors can operate in asymmetric mode for a long time without the actuation of protection systems, but with a significant deterioration in energy efficiency (Table 1.1). In its turn, at the asymmetry of current loading of the stator phases it results in essential increase in heating losses in separate phases of the machine at a small change in the value of total losses [37-40].

In addition, the acquired asymmetry may be caused by repairs without complete replacement of the windings. For example, in the event of damage to several sections of the IM stator winding due to short circuit, these sections are often not removed from the slots in order to reduce the amount of the repair work of high and medium power IMs, and the ends of the remaining sections are connected bypassing the faulty ones, i.e., damaged sections are excluded from the winding circuit. As a result of such operations, the stator winding becomes asymmetric (has different values of phase resistances and inductive reactances), and the motor further operates with such a winding [41-42].

Table 1.1. The possible degree of asymmetry of the stator windings at a damage in parallel branches

Number of poles	Maximum possible number of parallel branches in a phase	Asymmetry degree at the damage of one parallel branch, %	Asymmetry degree at the damage of one of elementary conductors in a parallel branch, %
$2p_n = 2$	a = 1,2	50	25-8.3
$2p_n = 4$	a = 1,2, 4	50; 25	25-8.3; 12.5-4.2
$2p_n = 6$	a = 1,2, 3, 6	50; 33.3; 16.7	25-8.3; 16.7-5.5;8.4-2.8
$2p_n = 8$	a = 1,2, 4, 8	50; 25; 12.5	25-8.3; 12.5-4.2; 6.3-2.1
$2p_n = 10$	a = 1,2, 5, 10	50; 20; 10	25-8.3; 10-3.3; 5-1.7
$2p_n = 12$	a = 1,2, 3,4, 6, 12	50; 33.3; 25; 16.7; 8.3	25-8.3; 16.7-5.5; 12.5-4.2; 8.4-2.8; 4.2-1.4
$2p_n = 14$	a = 1, 2, 7, 14	50; 14.3; 7.2	25-8.3; 7.2-2.4; 3.6-1.2
$2p_n = 16$	a = 1, 2, 4, 8, 16	50; 25; 12.5; 6.25	25-8.3; 12.5-4.2; 6.3-2.1; 3.1-1

A common type of IM asymmetry may also include the asymmetry of the magnetic system, which occurs due to the damage to the insulation between individual sheets of electrical steel and is manifested in a decrease in the value of magnetic induction [31].

In addition, a motor with the following faults is considered asymmetric: dynamic and static eccentricity of the rotor, mechanical weakening of the fasteners, incorrect mutual axial mounting of the active packages of the rotor and stator, electromagnetic defects of the rotor, breakage or contact in the rods or rings of rotor steel or only in the area of teeth) [43], stator defects (ellipse of internal boring of the stator relative to the axis of rotation of the rotor, weakening of the compression of the steel package) [44].

The described examples of IM asymmetry cause a significant deterioration in the efficiency of IM energy conversion and the emergence of variable components of active and reactive power in the motor [45]. Obviously, the variable component in IM instantaneous power leads to the appearance of a variable component in the electromagnetic torque and speed of the motor, increasing IM vibration and growth of power losses in the components of the machine. Such additional losses cause an increase in the overall temperature of the machine and local overheating [46].

Apparently, the influence of these factors results in a deterioration of the efficiency of both IM and VRED in general. In addition, even in the case of symmetrical exclusion of the same number of turns from all phase windings, which significantly reduces the reverse sequence current [39], the problem of the distortion of the shape of the torque curve is not solved. This is explained by the fact that the number of sources of distortion of the MMF curve around the stator increases, and there are variable components of instantaneous power. Besides, a change in IM performance (load capacity, maximum starting torque, etc.) is the consequence of such repairs. The inclusion of active filters in the ED circuit will not solve the problem of variable components of instantaneous power either, as such an approach is effective in terms of improving the operating conditions of the electrical network, but does not solve the problem of variable components of instantaneous power and electromagnetic torque of the load, in particular IM, if they are caused by its asymmetry.

The reliability of assessing the efficiency of the operation of VFED in general and IM in particular is the next factor that has a significant impact on ED efficiency and the effectiveness of the implementation of any energy-saving technologies. This provides a significant impact as the implementation of specific control modes for such EDs should be based on the reliable data on IM current efficiency and technical condition for the correct calculation of the necessary control effects. However, most of the existing methods for determining the performance indicators involve the use of additional equipment, which greatly complicates the necessary measurements without removing the motor from the process [47, 48]. A large number of methods do not allow taking into account the degree of asymmetry of the motor or change the modes of its operation [49-51]. The complexity of assessing VFED efficiency is also caused by distortions in the shape of the curves of voltages and currents and deviations of voltage and frequency values.

The sum of the fundamental harmonic and harmonics multiple of it and the switching frequency of the keys of the autonomous voltage inverter is the spectrum of the FC output voltage. Therefore, FC output voltage is polyharmonic, and FC is a polyharmonic voltage source for IM. The spectral composition of FC output voltage affects the efficiency of all ED elements [52, 53]. However, it should be noted that, when analyzing the mode of ED operation in general, it is necessary to take into account the IM "reverse" effect directly on FC. For example, the influence of IM asymmetry on the energy and thermal modes of operation of an autonomous inverter is currently not regulated. At the same time, the technical documentation for most industrial

SVI FCs indicates their ability to withstand a maximum current overload of 120-150% for a short time (about 60 s), and a 10% asymmetry of the IM stator results in a 130% current overload of individual phases. That is, IM asymmetry causes thermal overload of individual power switches. This is an additional factor in the need to develop and implement new methods for assessing and correcting the efficiency of EDs with asymmetric electromechanical converters.

In view of this, it is important to: *first*, develop a method for determining and assessing the current efficiency of ED, which would be adapted for the use in industrial conditions, without stopping the process, using minimal information about the object of the research and the minimum amount of required measuring equipment; and, *second*, develop a system to compensate for the impact of IM asymmetry by means of the VFED itself, without the installation of additional power equipment and without dismantling the motor.

1.3. The Analysis of the Prerequisites for Improving the Control Methods and Enhancing the Efficiency of VFED with Asymmetric IM

A number of researchers [52, 54-57] have shown that the possibilities of applying existing methods for assessing the efficiency of electric machines are significantly limited due to the increase in the number of AC drives operating at non-sinusoidal currents and voltages, and having a certain degree of initial or acquired design parametric asymmetry. Thus, the need to develop and implement new methods for assessing and correcting the efficiency of electromechanical converters is caused by the following reasons:

- the saturation of the fleet of industrial enterprise electric machines with motors with degraded energy characteristics, which are characterized by increased total losses and a significant level of variable components in instantaneous power. This is due to IM structural and parametric asymmetry caused by the obsolescence and wear of their main structural units in the process of long-term operation and overhaul;
- the absence of simple methods for assessing the efficiency of operation and recommendations for its improvement without

removing the IM from the technological process and the installation of additional equipment;

- the need for the diagnostics and monitoring of ED induction motors based on the analysis of energy parameters;
- the extensive possibilities of microprocessor and power converter technology, which allows expanding the functionality of variable-frequency ED.

1.4. The Analysis of Typical Damages to Power Electric Circuits of Variable-Frequency ED

The proper functioning of a variable-frequency ED depends on the reliable operation of the motor, three-phase SVI and control system. Each of these components may have failures in their operation, which can be divided as follows [59, 60]:

1. Electric motor malfunctions include:
   a. the malfunction of one or more phases of the IM stator;
   b. electric asymmetry;
   c. magnetic asymmetry;
   d. mechanical damages.
2. The malfunctions of the three-phase voltage inverter include:
   a. the malfunction of a single semiconductor key;
   b. the malfunction of one or two inverter arms;
   c. the malfunction or failure of the control driver.
3. The malfunctions of the control system include:
   a. the failure of the sensor of the angular rotation frequency;
   b. the failure of the current sensors;
   c. malfunctions in the formation of inverter control signals.

According to research [61], the most significant share of faults of variable-frequency ED is caused by malfunctions in the electric motor. It is known that IM malfunctions occur due to various damages and defects, which can be divided into external and internal ones [62, 63]. So the external malfunctions include: damage to the cables connecting the IM with the power supply; increasing the contact resistance of the motor terminals; malfunctions of the control equipment, reduced or increased mains voltage; IM overload; bad ventilation. IM internal malfunctions, in their turn, may be mechanical and electrical ones. Mechanical faults include damages to the bearing assembly, defects in the mounting of the actuator or gearbox to IM shaft, poor quality mounting of the IM to the base. Electrical damage, in turn, is divided into damage to the stator power circuit and rotor damage.

As shown in [65], about 46% of faults that reduce IM efficiency occur in power circuits, including both the phase windings of IM stator, and all conductors and connections between the power supply (energy converter) and IM.

The appearance of joints with high resistance (increased contact resistance of the motor terminals) can be identified among the most common IM external faults [65, 66]. The occurrence of such faults is associated with a number of factors such as high current or voltage, vibration, chemicals or pollution in the atmosphere, aging of the metal, high ambient temperatures, which may eventually initiate the mechanism of complete degradation of the electric machine. Increasing the contact resistance results in the growth of the contact temperature, which, in turn, leads to thermal expansion (and further weakening of mechanical stresses and slow changes in properties) and acceleration of oxidation or corrosion of contacts.

This problem is significantly complicated by the use of PWM FC, because the high frequency of IGBT switching leads to voltage peaks and high values of the derived voltage (rate of increase of electric potential) at the terminals of the motor. The length of the connecting cable between FC and IM and the high rate of the rise of the supply voltage of IM stator windings are the dominant factors influencing the occurrence of voltage peaks at the motor terminals [67]. The voltage signal applied to IM stator through the cable is partially reflected and leads to overvoltage, while excessively long cables increase the voltage throws at the terminals. In modern FC throws begin to be observed in cables several meters long and may reach the double value of the voltage in a DC bus at a length of less than 15 meters.

The appearance of high resistance junctions in electric machines in the early stages of development results in asymmetry of resistance of the stator windings. The operation in this mode causes an unbalanced supply voltage of IM, and, accordingly, an asymmetry of phase currents. As shown in [68], for every 1% of voltage imbalance, a current imbalance of 7% occurs, which leads to a braking torque (reverse sequence torque). Oscillations of the electromagnetic torque of the motor also occur in this mode, which results in pulsations of the angular frequency of rotation of the motor shaft and mechanical vibrations. In this case, long-term operation of the IM with high contact resistance of the terminals ultimately causes complete exclusion of the motor phase from the electric circuit.

In turn, damage to the phase windings of IM stator is the result of defects or unsatisfactory condition of their insulation. Thus, during IM operation, the insulation is the weakest and most vulnerable element of the structure, which is exposed to various influences, under the action of which its degradation and aging occur. In the general case, the effects that cause damage to insulation materials can be divided into several main groups [69].

1.4.1. Heat Loads

When an IM is designed, the service life of windings insulation is calculated depending on the temperature of its heating in a rated operation mode. However, if the insulation heating temperature exceeds the rated one, it causes a rapid reduction in its service life. Thus, in [70] it was shown that increasing the temperature by 10°C reduces the service life of insulation by about 50%.

The following can be identified among the main causes of IM excess temperature:

- fluctuations or asymmetry of supply voltage. As noted in [71], with phase voltage asymmetry at the level of 3.5%, the winding temperature with the highest current value increases by 25%;
- repeated starts of IM with short intervals;
- IM overload;
- damage to the cooling system;
- high ambient temperature.

1.4.2. Electrical Loads

During the operation, the insulation of electric machines is under operating voltage for a long time, and in addition it is periodically exposed to supply voltage higher than the rated one or a high value of the derived voltage (rapid voltage change at on/off, change of IM operation mode, FC power supply), which leads to accelerated wear of the dielectric material [72].

Electrical aging of insulation is very slow and gradually accelerates with the development of general destruction caused by various causes (thermal, mechanical, etc.). It is accompanied by delamination and loosening of insulation, the emergence of hollow spaces (air or gas layers), cracks. Various microdefects can, to some extent, be found in the new insulation, but as it ages, their number and size increase significantly. Ionization processes develop with the appearance of such inhomogeneities in the insulation, which is accompanied by its progressive destruction.

In insulation cavities, especially under overvoltage, there are partial discharges that destroy individual layers of insulation due to the thermal effect and mechanical splitting. In addition, the discharges are accompanied by chemical reactions with the formation of ozone, the interaction of which with air nitrogen and water vapor leads to the formation of nitric acid, which destroys the insulation.

Partial discharges sometimes cause so-called incomplete breakdowns, when only part of the insulation layers is broken. Over time, discharges become more frequent, and the voltage of their occurrence reduces. This process ends in complete failure. In addition to discharges in the thickness of the insulation under certain conditions, there are also surface discharges, which are accompanied by the destruction of the external layers of insulation. If no measures are taken to eliminate these phenomena, the insulation may fail within a few months.

1.4.3. Mechanical Loads

One of the most important factors of wear and aging of insulation is mechanical loads [73], which include static pressure on the insulation, bending and torsional forces, shocks and vibration. The source of mechanical effects includes electrodynamic forces, unbalanced rotating parts, centrifugal forces, pushes and shocks transmitted by drives or mechanisms.

The insulation of the slot part of the winding can feel compression under the action of electrodynamic forces, and in the presence of gaps in the slot, it is also prone to shocks and abrasions against the walls of the slot. When the front parts of the winding bend, the greatest stresses occur at the exit of the rods or coils from the slots, where the insulation is subjected to compressive and tensile stresses. In addition, it crumples on the gaskets and in contact with the bandages.

In most cases, the mechanical effects are cyclical, alternating in nature, and vibration with a frequency of 100 Hz is most typical [74]. Vibration amplitudes grow tenfold periodically in transients (start, reverse, short circuit) due to the increased current in the windings and the quadratic dependence of electrodynamic forces on current. Thus, research [72] shows that vibration reduces the service life of insulation several times.

The mechanical characteristics of the insulation significantly depend on the temperature. As it heats up, the strength of the insulation decreases rapidly, and at the same time the insulation becomes more elastic. Modern types of insulation have significant resistance to static loads. However, even with relatively small deformations, there is a significant reduction in breakdown voltage, which occurs approximately according to the linear law. For each temperature there is a certain limit of deformation, beyond which the removal of the load does not lead to the restoration of the primary dielectric properties. This indicates that significant deformations are accompanied by the appearance of irreversible structural changes in the form of cracks, breaks, delamination.

1.4.4. The influence of Environmental Factors

High humidity, high concentration of harmful chemicals and foreign substances in the environment, natural aging of the insulation can also lead to the accelerated failure of IM stator windings. So humidification of windings occurs in case of long breaks in the motor operation, at direct getting of water or steam in it as a result of storage of the motor in a damp unheated room. The intrusion of foreign substances into the motor may gradually cause a short circuit between the phases of the windings and the housing.

In turn, the presence of foreign substances outside the motor can also adversely affect its operation, as excessive contamination causes a reduction in heat dissipation (which leads to an increase in operating temperature, thereby reducing the service life of the insulation) and premature failure of bearings due to high localized loads.

In some cases, damage to the insulation may not be the result of aging or wear. The causes of such damage may include *design defects* and *defects in production technology*. In the general case, the design defects may include [75]:

- *high coefficient of filling the slots* (an increase in the number of conductors in a slot causes application of considerable efforts at laying of a winding that leads to local damages to the insulation and turn-to-turn short circuits in the course of operation);

- *small margin of heat resistance of insulation* (in operating conditions IM is periodically subjected to various overloads caused by malfunction of the drive mechanism or forcing its mode of operation, deviations from frequency or voltage, temporary increase in ambient temperature. Overloads are accompanied by rising winding temperature, leading to rapid aging of the insulation);
- *small air gap* (slight deformation of the seating surfaces, bearing wear, bending of the shaft lead to a significant uneven air gap, which causes vibration, and in the most severe cases – contacting of the rotor and the stator).

In turn, the defects of manufacturing technology are caused by the internal structure of electric motors, which is determined by the interaction of parts with significantly different properties [76]. The mechanical strength of electrical insulation is much lower than the strength of the surrounding parts, so it is easily damaged by deformation and impact, in contact with sharp edges of metal parts, when foreign substances enter it. The destruction of the winding insulation is also increased by the presence of burrs formed at the edges of the cuttings and, in particular, in the slots. Also, a high probability of insulation damage occurs at high speeds of winding the wire, or when laying the latter in the slot. Such local defects often develop relatively quickly and lead to insulation breakdown long before a significant deterioration of its properties due to electrical or thermal oxidative damage.

The considered factors cause various damages to the windings of IM stator phases (Figure 1.1). Usually these damages develop gradually, but eventually lead to complete failure of the motor. A short circuit of the motor phase to ground is the most significant damage. Other damages may include: turn-to-turn short circuits, short circuits between phase coils, interphase short circuits, breaks of parallel branches of phases or stator phase break.

Interturn damage to the stator windings is the result of short circuits of two or more turns in one IM phase. In the early stages of development, turn-to-turn short circuits do not significantly affect the motor performance, but cause asymmetry of electromagnetic parameters (resistance, scattering inductance and mutual inductance) in motor phases [77]. In short-circuited turns the current flux is much higher than the operating one and, therefore, the winding temperature increases, which may result in phase failure, parallel section or phase or interphase short circuit. Studies [70, 78] show that the share of IM failures due to short circuits is about 30-40% of all failures.

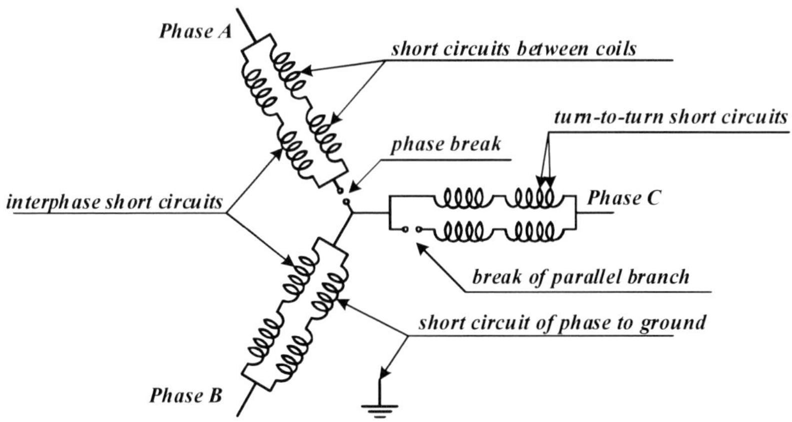

Figure 1.1. The damages of IM stator.

The occurrence of turn-to-turn short circuits is enhanced by the use of FC with PWM voltage [79]. In this case, the motor windings receive pulse voltage with significant overvoltage peaks, the total value of which exceeds the amplitude of IM rated voltage. The greatest electrical load falls on the turn-to-turn insulation of the first turns, which are electrically located closest to the junction of the power cable and winding [80]. In addition, there are noticeable dielectric losses in the insulation of the machine with a rapid increase in supply voltage. All this leads to rapid aging of the insulation, which affects IM reliability and service life. As a result of these processes, the service life of IM insulation is significantly reduced.

The break of IM stator phase is the most significant consequence of the appearance of turn-to-turn circuits. It is known that when the phase is broken, IM starting torque is zero, and when the motor is turned on, a short circuit mode takes place, the currents in serviceable phase windings significantly exceed the rated ones, which causes intense overheating and rapid failure of insulating materials. If a break occurs during the operation, IM may continue to work, but the currents in the operating phase windings significantly exceed the rated ones, which leads to overheating of the insulating materials of the windings and motor failure.

Thus, as a result of any malfunction of the stator windings the motor operates in an asymmetric mode, which leads to the redistribution of currents in the phase windings and to changes in electromagnetic parameters in IM phases. At the same time, the values of currents considerably exceed the rated ones in separate phases (parallel branches of phases, short-circuited turns, etc.). It causes increased heat losses and an excess of the temperature of insulating materials of the maximum admissible values.

Asymmetric current loading of IM phases results in local overheating in some FC semiconductor switches. Due to the fact that the energy of losses in power switches is released in a semiconductor crystal and dissipated in the form of heat, FC long operation in this mode can lead to significant overheating of individual switches, and subsequently to their complete failure [81, 82].

1.5. The Review of the Existing Methods for the Control of IM with a Damage to the Stator Power Circuit

The constantly growing trend of expanding the scope of the use of variable-frequency EDs with IM in industry makes the issue of their reliability and fault tolerance rather topical. Breakdowns or failures can cause unplanned downtime of the process, which leads to significant economic losses. These losses often exceed the cost of ED maintaining [83]. Accordingly, fault-tolerant control (FTC) systems are of particular interest today. They are able to detect various types of ED damage at the initial stage and quickly adapt the control law so as to maintain the specified parameters (quality of production, safety, etc.) [84].

The research presented in [85] reveals that to achieve this purpose, the control system must be built in such a way that it can use the optimal control method on the basis of the available feedback signals. Thus, the FTC system consists of two parts: a fault detection system and a strategy to neutralize it. While the former part monitors the condition of system components such as sensors, motor, inverter, etc., the latter chooses the optimal control strategy.

The analysis of damage in the power circuit of IM stator makes it possible to distinguish two main cases of non-stationary modes of operation of variable-frequency EDs:

- *operation with minor defects or damage in the early stages of development.* In this mode, the efficiency of the system is not lost, but there is a loss of control quality, a decrease in energy efficiency of the electromechanical energy conversion process, a significant increase of losses and variable components of electromagnetic torque and active power consumption;

- *operation with significant damage.* In this mode the IM energy performance is significantly deteriorated, the critical torque of the motor and accordingly the maximum power transmitted to the shaft are reduced. In the signal of the electromagnetic moment of the motor there are fluctuations with a frequency equal to twice the frequency of the supply voltage. IM long operation in this mode ultimately leads to emergencies.

In general, FTC can be divided into two types: passive and active. Examples of passive FTCs include systems in which the main control is switched to the backup control type when certain damage is identified. Thus, when vector control with a speed sensor is used as the main mode, the transition to the mode of sensorless vector or scalar control takes place.

Unlike passive systems, active FTCs are able to respond to failures of VRED components by changing control effects, while maintaining stability and acceptable performance of the entire system. First, the fault is identified and localized, and then the control law is changed to achieve the desired result. In such systems, the consequences of faults are compensated for by selecting a pre-calculated or synthesized new control law.

At present, a significant number of methods of forming control and compensatory effects are used for IM FTC [86]. The analysis of literature makes it possible to create a classification of the most widespread methods used in FTC of IM as part of VFED.

Depending on the nature of the interaction with the object of control:

- active [87-91, 94-98, 61, 100-105, 110-124];
- passive [92, 99, 106-109].

Depending on the type of frequency control:

- scalar [92, 95, 98, 99, 101, 106, 116-120];
- vector [87-91, 93, 94, 96, 97, 100, 102-106, 110, 112-114, 116, 121-124];
- direct torque control (DTC) [111, 114].

By the nature of the damage to VFED elements:

- break of the stator phase [87-100, 108, 109, 112-115, 117-124];
- asymmetry of the resistances of the stator windings [108-114, 116-124];
- rotor eccentricity [117-120];
- appearance of joints with high resistance [101-105];
- breakage of rotor rod or change of rotor resistance [106, 176];
- failure of the elements of the power energy converter [106-108].

Depending on the selected coordinate system:

- three-phase U V W [92, 95, 98, 99, 117-120, 121-124];
- fixed α β [87-91, 97];
- rotating d q [87-91, 93, 94, 96, 97, 100, 102-105, 112, 113, 116, 121, 122].

Depending on the selected parameter, the components of which are adjusted or compensated for:

- electromagnetic torque [87-91, 95, 97, 98, 114, 115, 122];
- electric power at the input [117-121];
- stator currents [92, 99, 101-105, 112, 113, 115, 121-124].

Depending on the methods of obtaining the source information and damage diagnostics:

- the use of an observer based on a bilinear model of a defective IM [116];
- the determination of damage and eccentricity of the rotor using wavelet analysis [109];
- the application of rotor flux observer based on "backstepping" method [113];
- the use of a neural network to determine the degree of asymmetry of the stator windings [114];
- the determination of IM damage by the analysis of current harmonics or power consumption [117-120];

- the determination of damages by the size of the parameter variable components [93, 94, 96, 100];
- the calculation of rotor resistance using the extended Kalman filter [91];
- the calculation of electromagnetic torque through direct measurement of a field in the gap and phase currents, or through measurements of phase currents and voltages [92, 99, 101-105, 121].

Depending on the methods for the calculation and formation of corrective actions:

- the displacement of the current vectors of the working phases by angle $\pi/3$ [92, 99];
- the introduction of additional voltage harmonics to neutralize the second and higher harmonics of the torque pulsations [95, 98];
- the application of additional control circuits with classical PI controllers [102-105];
- the introduction of a variable component into the control circuit of the torque-generating component of the current [112-116, 122];
- the use of instantaneous power theories: modified p-q power theory method [121], cross-vector power theory [117-120];
- the separate correction of flux linkage in three phases [123-124];
- the use of an additional block of coordinate transformations, the mathematical apparatus of which is based on certain positions of the current vector [93, 94, 96, 100];
- the use of a genetic algorithm to adjust the coefficients of the angular rotation frequency regulator [87-91, 97];
- switching between control algorithms from more complex to simpler ones [106-109].

The performed analysis revealed a large number of existing methods for the control of IMs with damages, which to some extent ensure performance and allow correcting the modes of operation of the motor as part of a variable-frequency ED.

Today, only passive methods are widely used. They imply a switch from more complex control algorithms to simpler ones when various damages occur. They are widely used in the event of damage to the information part of the regulated ED: the failure or malfunction of the angular rotation frequency or the current sensors.

Conclusion

1. The analysis of the known methods for induction motor control revealed the presence of a large number of different control algorithms and allowed identifying the most commonly used and promising methods for their further research.
2. Based on the analysis of the design features and damage statistics of induction motors, it is shown that the magnetic and electrical asymmetry of the stator windings, resulting from electrical damage such as turn-to-turn short circuits, the break of the parallel section of the stator phase winding, the break of one elementary conductor in the winding, is the most common cause of occurrence of the variable components of power and electromagnetic torque.
3. The analysis of the features of operation and control of the variable-frequency electric drive demonstrated the possibility of expanding the scope of the frequency converter, which can act for IM not only as a power source but also as a compensator, i.e., as a source of compensation current to eliminate the asymmetry of the stator windings.
4. Using the classification proposed by the authors makes it possible to choose the most rational methods and means of fault-tolerant control of induction motors as part of a variable-frequency electric drive, taking into account the hardware and software capabilities of modern control systems.

Chapter 2

Mathematical Modeling of the Systems of a Variable-Frequency Electric Drive

2.1. The Implementation of the Power Unit of a Variable-Frequency AC Electric Drive with a Frequency Converter

The problem of induction and synchronous motors control in the systems of variable-frequency AC ED is solved by using frequency converters (FC). Electric drives with frequency vector control are not inferior to DC electric drives in terms of their regulating properties, and in terms of economic and small-size indicators they mostly predominate [2]. The use of FC with AC motors in automated electric drive systems provides such indicators as the efficiency of operation with the simultaneous possibility of a wide range of angular rotation frequency and loads, wide range of angular rotation frequency control both down and up from the rated one, increased value of the overload capacity at the torque and high speed in the regulation of the parameters (current, rotation frequency and position of the IM rotor).

The circuits of FC power part are very diverse, and the use of a particular circuit is determined by the specific requirements for the electric drive system. At the same time, frequency converters with a direct current link became widespread [125] (Figure 2.1).

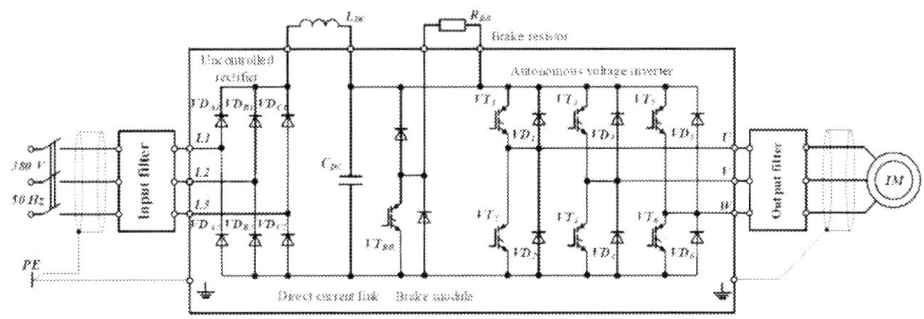

Figure 2.1. A circuit of the power part of the variable-frequency electric drive with a frequency converter.

A controlled or uncontrolled rectifier can be installed at the input of the power channel of this type of converters. The use of a fully controlled thyristor or transistor rectifier provides two-way energy exchange between the network and the motor, the consumption of almost sinusoidal current from the network, the ability of the adjustment within a wide range of the power factor of the converter. However, frequency converters with an uncontrolled rectifier are most commonly used today. In this case, in low-power converters a single-phase bridge rectifier is most often used, and in high-power converters – a three-phase bridge rectifier.

The AC energy with constant voltage and frequency values in a FC with a DC link is first converted into DC energy. The energy of direct current enters the input of the inverter and is again converted into the energy of three-phase alternating current, but with adjustable voltage and frequency parameters. Thus, a double conversion of energy takes place in the frequency converters of this type, which somewhat reduces its efficiency, but other significant advantages of these converters provide them with a dominant position in the modern automated electric drive.

An energy storage device, which is a *LC* filter, is installed between the rectifier and the inverter. In this case, the inductance in the DC circuit of the converter smoothes the pulsation of the rectified current and limits the rate of increase of the emergency current through the power switches with a short circuit on the side of the rectified current. *LC* filter capacitor is designed to short the reactive component of the stator current.

The parameters of the said filter are simply defined as follows: capacitor capacity C_{DC}

value of the inductance is determined by expression:

$$L_{DC} = \frac{s+1}{m^2 \omega^2 C_{DC}}, \tag{2.1}$$

where s – pulsation smoothing coefficient ($s = q_{\text{in}} / q_{\text{out}}$, q_{in} – pulsation coefficient at the filter input; q_{out} – pulsation coefficient at the filter output, is taken within 0.01–0.1); m – the number of output voltage pulsations for the period; ω – supply voltage frequency.

The DC link also usually includes a braking unit, which contains a power transistor switch and a load resistor. Low-power braking units are devices built into FC, and high-power braking units are performed as separate devices connected to DC buses. Braking units are used when energy recovery occurs infrequently and for a short period of time, for example, when braking the drive to stop. For long-term braking, controlled rectifiers are used instead of an uncontrolled rectifier, or recovery units are used, which are additionally connected to DC busbars.

The output parameters of voltage (amplitude and frequency) of FC with a DC link are adjusted by the inverter operating in the mode of pulse-width modulation (PWM). The use of pulse-width method of output voltage has a number of advantages, which are manifested in the following [126]:

- the shape of the output current is significantly closer to sinusoidal, the uniformity of rotation of the motor shaft improves, the range of adjustment of the angular rotation frequency expands;
- the speed of the electric drive considerably increases as the power filter actually does not influence the channel of regulation of output voltage of the converter;
- the power coefficient at the output of the inverter significantly improves.

However, when using converters with PWM, it is necessary to take into account a number of negative side effects, namely [126]:

- the broadband spectrum of electromagnetic interference emitted by the motor power lines from the inverter, and propagated through the cables of the FC power supply network;
- the generation of capacitive currents in power cables and motor design elements due to steep fronts and high frequency of the modulating component of the inverter output voltage;
- pulse voltage with significant overvoltage peaks is applied to the motor windings, the total value of which exceeds the amplitude of IM rated supply voltage, which can cause a breakdown of the insulation of the motor windings, especially during its long operation;
- the creation of additional modulated motor noise due to modulating high-frequency oscillations of magnetic induction in the magnetic circuits of the motor, magneto-strictive conversion of these oscillations into mechanical, then acoustic vibrations;
- high-frequency pulsations of IM electromagnetic moment, which are caused by the modulation component of the inverter voltage;

- at low PWM frequencies high-frequency pulsations of the motor current can reach up to 5-10% of the rated current, which leads to additional heating of the motor.

The manifestation of these negative effects can be significantly limited by equipping the output of the inverter with filters that reduce the modulation component of its output voltage and impulse overvoltages at the load terminals [127]. So, if the length of the power cable exceeds 50–100 m, to limit overvoltages on the terminals of the drive motor and protect the insulation of its windings from breakdown, it is recommended, in close proximity to the motor, to connect an output three-phase choke or a capacitive filter and a choke that provide the following:

- the reduction of high-frequency harmonics in the motor current;
- the limitation of the amplitude and rate of increase of the emergency short-circuit currents, which leads to an increase in the time to reach the maximum short-circuit current, thereby providing the necessary time for the operation of FC protection systems;
- the compensation for capacitive currents of long motor cables that prevents formation of big capacitive currents and accordingly interferes with false operation of FC protection systems;
- the reduction of the overvoltage on motor windings.

The main requirement for this filter is to provide a given coefficient of alternating voltage harmonics in the steady state. Thus, the recommended value of the filter inductance is determined by expression:

$$L \geq \frac{0.25E_{max}}{(0.4..0.6)f_{pwm}I_n}, \tag{2.2}$$

where: $E_{\max}$ – the maximum voltage of the direct current link; f_{pwm} – PWM frequency; I_n – the rated current of the motor.

The capacity of the filter is determined by expression:

$$C = T_{pwm}^2 / 64kL \tag{2.3}$$

where: T_{pwm} – PW frequency period ($T_{pwm} = 1/f_{pwm}$), k – the coefficient of higher harmonics.

The output chokes are connected in each phase in series with the motor windings and the capacitors are connected in a triangle in parallel to the motor. Accordingly, the capacitors do not significantly affect the overall resistance at FC output.

In turn, the reduction of these interferences in the power supply network is achieved by installing filters in FC input circuit [128]. Typically, such a filter is represented by a three-phase choke, which is designed to reduce the current harmonics generated by the rectification/recovery unit. The effect of the choke depends on the ratio of the short-circuit power of the supply network to the drive power; the recommended value of this ratio is more than 33: 1. The input choke also allows reducing the current emissions caused by power surges in the supply network.

The frequency of modulation of the output voltage is an important parameter in the analysis of the modes of operation of FC with a direct current link. The optimal choice of this parameter is essentially a variational problem, the content of which is determined by the following conflicting factors [129]. So increasing the frequency of PWM allows getting a number of positive effects:

- increasing the dynamic accuracy of reproduction of input set influences by pulse-width modulators;
- expanding the operating frequency range of FC-IM system;
- reducing the amplitude of modulation pulsations of currents, flux linkages and electromagnetic torque of the motor, as well as dependent components of modulation losses in the motor and power supply circuit;
- reducing the additional modulated motor noise, which, in some cases, makes it possible to abandon the installation of output filters;
- creating the conditions for speed increase and improvement of other indicators of the quality of the closed systems of automatic regulation of AC electric drives and their use for the control of technological processes with the increased requirements for the electric drive.

However, the growth of PWM frequency increases a number of negative effects, namely:

- due to the growth in the switching frequency of the power switches, the switching losses in the inverter increase proportionally and its allowable useful power decreases;
- the effective values of capacitive currents in power cables and the elements of the motor design grow, which increases the losses from these currents in the FC-IM system.

In turn, reducing the PWM frequency has the opposite effect, namely: switching losses are reduced and the useful power of the inverter is increased, the losses from capacitive currents are reduced. However, this decreases the range of operating frequencies, increases the pulsations of current, flux linkages and electromagnetic torque, heightens the modulation losses and the level of acoustic noise during motor operation.

Thus, the optimal value of the modulation frequency is selected in each case taking into account the specific conditions of the use of the converter and the necessary requirements for the system of variable-frequency ED. In modern ED with a power of 0.75 to 30 kW, the PWM frequency is 1–16 kHz, for ED with a power of more than 30 kW, the PWM frequency is 1–8 kHz. Thus for more powerful ED smaller value of PWM is chosen.

In addition to the above, the negative effects in the operation of FC with output voltage PWM include the impact of the dynamic properties of imperfect power switches on ED output characteristics. These properties may include the delay of on and off, "dead zone" in switching upper and lower inverter phase switches [130]. In this case the following may be observed: the distortion of the shape of the output voltage of the inverter, which leads to additional pulsations in the current, torque and angular speed; the underutilization of the inverter power supply voltage; the possibility of intermittent currents. This effect is especially strong at high modulation frequencies and when the drive is operating in the low voltage range (at low angular speeds). In order to minimize the impact of the "dead zone" in control systems, algorithms are used to compensate for it.

In general, the following advantages of converters with a DC link can be identified:

- the ability to obtain a wide range of frequencies at the output of the converter, regardless of the frequency of the power supply;
- the possibility of using relatively simple power circuits and control systems;

- the ability to implement a variety of control algorithms that meet the requirements for automated ED systems.

The main disadvantages of FC with a DC link include:

- double energy conversion, which increases energy losses, degrades the mass and size of the converter;
- the presence in the DC link of a power filter containing a battery of capacitors of large capacity or a reactor with significant inductance, which leads to a deterioration in mass and size, and the reliability of electrolytic capacitors reduce.

To adequately take into account all these features of FC with a DC link and PWM voltage in the research using mathematical models, it is necessary to use specialized packages, such as: Workbench, Multisim, DesignLab, Micro-Cap, OrCAD, MATLAB SimPower Systems, PSIM.

2.2. Mathematical Modeling of Alternating Current Electric Drive Systems with Vector Control

IM vector control systems are based on the representation of physical variables of the motor by spatial vectors, in which both the module and the position in space can change. The fundamental principle of vector control is the orientation of ED vector variables relative to each other [131, 132]. The orientation can be performed on almost any vector variable, but usually one chooses the variables, the orientation of which allows getting the best dynamic and static properties of the ED and the simplest structure of the control system.

A mathematical description of the IM in a rotating coordinate system, oriented to the resulting vector of the flux linkage of the rotor, is the basis for the implementation of vector control systems (Figure 2.2).

This approach is often called field orientation method. In this case, IM acquires characteristics close to those of a DC motor. The vector-controlled ED provides separate regulation of the magnetic flux and electromagnetic torque of the motor; and in the mode of maintaining the constant flux linkage of the rotor the maximum allowable speed in torque control is obtained.

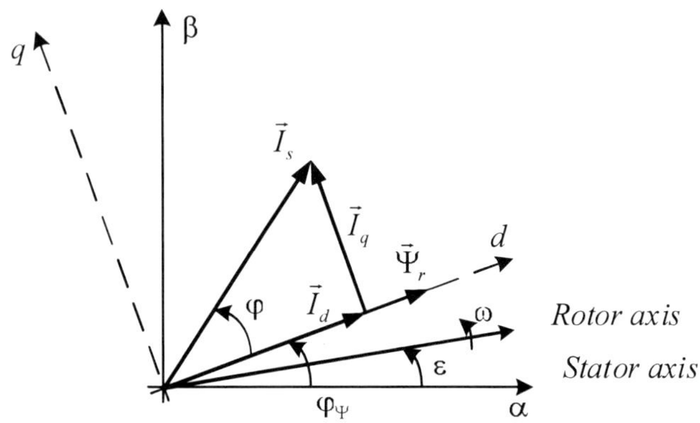

Figure 2.2. Orientation by the rotor flux linkage vector.

Compared to scalar control systems, IM vector control has a number of advantages: high accuracy and wide range of rotation frequency control of the IM shaft; rapid response to load changes – sharp load jumps are practically not reflected at ED angular rotation frequency; the ability to maintain the nominal value of the torque on the shaft even at zero rotation frequency; the possibility of implementing energy-saving modes of operation; high motor efficiency due to reduced losses caused by magnetization and heating.

Despite the obvious advantages the vector control method has certain disadvantages: great complexity of calculations in the control system; knowledge of IM parameters is necessary for work; impossibility of the use in group ED systems.

The system of differential equations describing IM electromagnetic and electromechanical processes, written with respect to the flux linkage of the rotor and the stator current in the rotating coordinate system (*d, q*), oriented by the rotor flux linkage vector, is of the following form:

$$\begin{cases} \sigma L_s \dfrac{dI_d}{dt} = -R_s I_d + U_d + \sigma L_s \omega_\Psi I_q - k_r \dfrac{d\Psi_r}{dt}; \\ \sigma L_s \dfrac{dI_q}{dt} = -R_s I_q + U_q - \sigma L_s \omega_\Psi I_d - k_r \omega_\Psi \Psi_r; \\ T_r \dfrac{d\Psi_r}{dt} = -\Psi_r + L_\mu I_d; \\ \omega_\Psi = \omega_e + \omega_2 = \omega_e + \dfrac{L_\mu}{T_r}\dfrac{I_q}{\Psi_r}; \\ M = \dfrac{3}{2} p_n k_r \Psi_r I_q, \end{cases} \quad (2.4)$$

where R_s, R_r – resistances of stator and rotor phases; L_s, L_r, L_μ – inductance of stator, rotor phases and mutual inductance; $T_r = L_r / R_r$ – rotor circuit time constant; $k_r = L_\mu / L_r$; σ – scattering coefficient; U_d, U_q, I_d, I_q – projections of stator voltage and current vectors on axes *d* and *q*; Ψ_r – rotor flux linkage vector module; ω_Ψ – the rotation frequency of the rotor flux linkage vector; $\omega_e = p_n \omega_r$ – electric rotation frequency of the rotor; ω_2 – slip frequency; p_n – the number of IM pole pairs.

Based on the above equations and neglecting the internal cross-links, which can be compensated for by feedback to the input of the frequency converter, the block diagram of the power channel of the FC-AD system takes the form shown in Figure 2.3. The indication in the Figure includes: R_e – total resistance of the motor ($R_e = R_s + k_r^2 R_r$); T_S – the equivalent time constant of the induction motor ($T_S = \sigma L_S / R_e$).

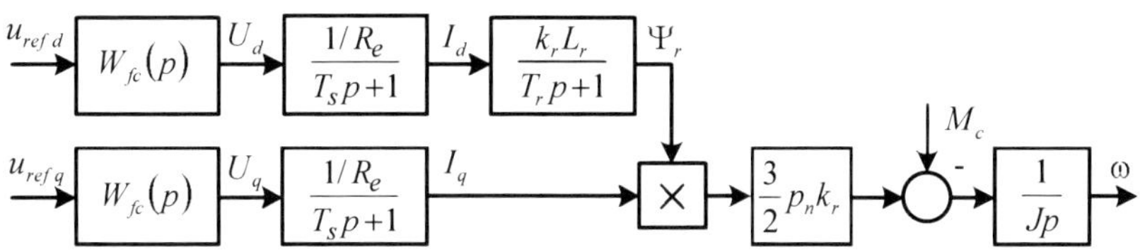

Figure 2.3. The block diagram of the FC-IM power channel in the rotating coordinate system oriented by the rotor flux linkage vector and at the compensation for the internal cross-links.

The given block diagram of ED with vector control in a rotating coordinate system uses projections of the corresponding spatial vectors on the corresponding axes as input and output parameters. These parameters are the values of direct current, which allows creating a vector control system similar to DC ED control system [133, 134].

Improving the dynamic properties of ED with IM in vector control is the result of the fact that in transient modes it is possible to maintain a constant flux linkage of the rotor in contrast to scalar control, where the rotor flux linkage in transient modes changes when stator and rotor currents vary, which results in the reduction of the rate of the change of the electromagnetic torque and occurrence of its oscillations [135]. In the vector-controlled ED, the rate of the change of the electromagnetic torque corresponds to the rate of the change of the torque-generating component of the stator current I_q, similarly to the change of the torque when the armature current changes in a DC machine.

In a real system with three-phase IM, when an ED control system based on the system of equations 2.4 is created, the control system should include coordinate converters that perform direct and inverse conversion of DC values from a rotating coordinate system into a three-phase value system in a fixed coordinate system [136]. The block diagram of the conversion of electrical values from one coordinate system to the other is presented in Figure 2.4.

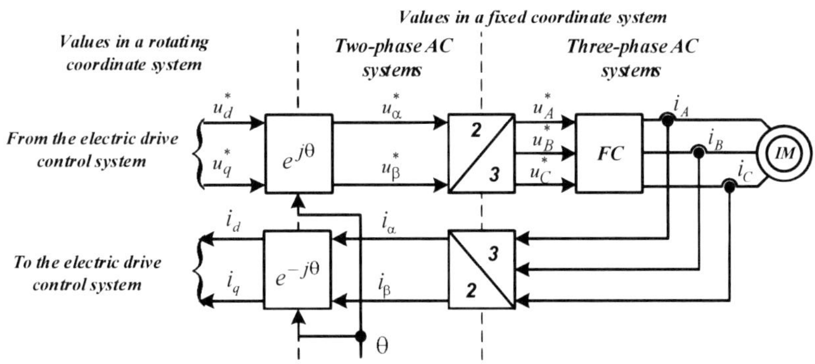

Figure 2.4. Coordinate transformation in vector control.

The transformations are carried out in two stages. In the feedback channel, the three-phase system of periodic current values is converted into a two-phase one, and then into the projection of the spatial vector on the axis of the rotating coordinate system, which is DC signals. In a direct channel, a two-phase system of variables is first formed from DC signals, and then it is transformed into a three-phase system of periodic values. The Clarke and Park transformations are used to implement these transformations in IM vector control system.

Clarke's transformation transforms a three-phase system of periodic currents i_A, i_B, i_C into a two-phase orthogonal system of periodic currents in axes i_α, i_β.

$$\begin{bmatrix} i_\alpha \\ i_\beta \\ i_\gamma \end{bmatrix} = \begin{bmatrix} 2/3 & -1/3 & -1/3 \\ 0 & 1/\sqrt{3} & -1/\sqrt{3} \\ 1/3 & 1/3 & 1/3 \end{bmatrix} \cdot \begin{bmatrix} i_A \\ i_B \\ i_C \end{bmatrix} \tag{2.5}$$

The obtained orthogonal current signals in the stator fixed coordinate system are converted into current components i_d, i_q in the rotating coordinate system d, q according to the Park transformation:

$$\begin{bmatrix} i_d \\ i_q \\ i_0 \end{bmatrix} = \begin{bmatrix} \cos(\theta) & \sin(\theta) & 0 \\ -\sin(\theta) & \cos(\theta) & 0 \\ 0 & 0 & 1 \end{bmatrix} \cdot \begin{bmatrix} i_\alpha \\ i_\beta \\ i_\gamma \end{bmatrix} \tag{2.6}$$

The inverse Park transformation allows the transformation from a rotating coordinate system d, q to a fixed one α, β:

$$\begin{bmatrix} u_\alpha \\ u_\beta \\ u_\gamma \end{bmatrix} = \begin{bmatrix} \cos(\theta) & -\sin(\theta) & 0 \\ \sin(\theta) & \sin(\theta) & 0 \\ 0 & 0 & 1 \end{bmatrix} \cdot \begin{bmatrix} u_d \\ uq \\ u_0 \end{bmatrix} \tag{2.7}$$

The inverse Clarke transformation provides a transition from a two-phase orthogonal coordinate system α, β to a three-phase system A, B, C:

$$\begin{bmatrix} u_A \\ u_B \\ u_C \end{bmatrix} = \begin{bmatrix} 1 & 0 & 1 \\ -1/2 & \sqrt{3}/2 & 1 \\ -1/2 & -\sqrt{3}/2 & 1 \end{bmatrix} \cdot \begin{bmatrix} u_\alpha \\ u_\beta \\ u_\gamma \end{bmatrix} \tag{2.8}$$

Only the channel of the stator windings is used in the vector control of the IM with a short-circuited rotor. The currents form a magnetic field and determine the torque. It is necessary to control the module and the position of the stator current vector relative to the rotor flux linkage vector (Figure 2.2). This comes down to controlling the amplitude and phase of the stator current.

To organize such a control, it is necessary to have information about the modules and positions of the flux linkage vector and the load moment on the motor shaft, which requires the presence of flux sensors installed in the air gap of the machine and the angular rotation frequency sensor on the motor shaft [137]. In this configuration, the vector-controlled ED can be compared with the DC ED.

As a part of general industrial IM, the mentioned sensors are absent as their introduction leads to the complication of the motor design and an essential increase in its cost. Modern vector control technologies can circumvent these difficulties by using an IM adaptive model [138]. In this case the ED control system must measure the values of currents and voltages at the output of the converter with high accuracy, calculate the coordinates of the IM at high speed, accurately model the processes in the motor and ultimately calculate FC control signals.

The functional diagram of the variable-frequency ED with vector control of the IM with the orientation by the rotor flux linkage vector is shown in Figure 2.5.

The adjustment system is made in a rotating coordinate system. Coordinate conversion in the forward channel and in the feedback channel is carried out according to the rules of coordinate conversion in the vector control system (equations 2.5-2.8).

For the direct and inverse conversion of variables from a rotating coordinate system to a fixed coordinate system it is necessary to have information about function $\sin\theta$ and $\cos\theta$, where θ – the angle between axis α of the fixed coordinate system and axis d of the rotating coordinate system. Related calculations, as well as the determination of the modulus of the rotor flux linkage vector, which closes the flux control circuit, are implemented by expressions:

$$\hat{\omega}_e = \omega_e + \frac{L_\mu}{T_r \Psi_r} I_q; \qquad \Psi_r = \frac{L_\mu}{T_r p + 1} I_d; \qquad \theta = \int \hat{\omega}_e dt, \qquad (2.9)$$

where $\hat{\omega}_e$ – the current value of the voltage frequency on the stator of the motor.

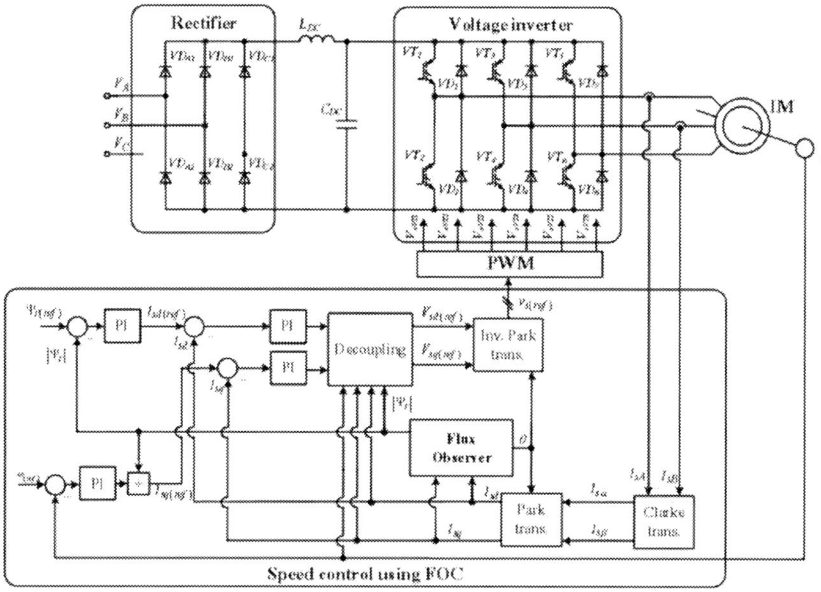

Figure 2.5. The block diagram of variable-frequency ED with vector control with the orientation by the rotor flux linkage vector.

To provide independent control of the flux linkage of the rotor and the electromagnetic moment (angular rotation frequency) of the IM, it is necessary to exclude the mutual influence of the projections of the stator current vector I_d and I_q in the rotating coordinate system, which is characterized by the presence of cross-links according to these projections in the block diagram [139, 140]. This is solved with the help of the compensation unit, where the cross-links are compensated for by introduction of the same cross-links as in IM structure, into its input, but taken with opposite signs:

$$\begin{cases} U_{kd} = -\dfrac{1}{k_{fc}} I_d \sigma L_s p_n \omega_r; \\ U_{kq} = \dfrac{1}{k_{fc}} \left(I_q \sigma L_s p_n \omega_r + p_n \omega_r \Psi_r \left(\dfrac{L_\mu}{L_r} \right) \right), \end{cases} \tag{2.10}$$

where U_{kd}, U_{kq} – compensation voltage; k_{fc} – FC transmission coefficient.

The vector control system is based on the principles of subordinate regulation. It contains two independent external feedback circuits: the angular rotation frequency and flux linkage of the rotor, and two subordinate circuits by the stator current component, creating a negative connection by the current vector [141, 142]. Rotation frequency information is supplied from the output of the angular rotation frequency sensor. After subtracting it from the specified value of the angular rotation frequency, the received error signal is fed to the angular rotation frequency controller, at the output of which the signal of the torque is formed. Dividing the torque reference signal into the flux linkage module of the rotor makes it possible to obtain the torque signal of the torque-generating component of the stator current I_q. The flux linkage is stabilized by a flux regulator (FR), which generates a reference signal of the flux-forming component of the stator current I_d.

The control algorithm can be synthesized in any coordinate system due to the fundamental property of linearization by the IM model feedback. However, the concept of IM control with machine field orientation, as a method of providing independent control of the flux and torque of IM, is more clearly represented in the rotating coordinate system (*d, q*).

Usually in vector control systems the regulators of the torque-generating I_q and flux-forming I_d components of the stator current, as well as the regulator of the flux linkage of the rotor are synthesized to the modular optimum, and the angular frequency controller is synthesized to the symmetrical optimum. In this case a simplified block diagram of IM mathematical model in vector control is used (Figure 2.3). Then the rotor flux linkage control channel includes a flux linkage regulator, a current regulator and a first-order inertial link frequency converter. The rotor angular rotation frequency control channel includes an angular rotation frequency controller, a current regulator and a first-order inertial link frequency converter.

The parameters of the regulators in the vector control system can be determined as follows:

- the transfer function of the current regulator:

$$W_{cr}(p)=\frac{(T_s p+1)R_e}{2(T_{\mu c}+T_c^s)k_{fc}k_c^s}=k_{pc}+k_{ic}\frac{1}{p}, \tag{2.11}$$

where $k_{pc} = \dfrac{R_e T_s}{2\left(T_{\mu c} + T_c^s\right)k_{fc}k_c^s}$; $k_{ic} = \dfrac{R_e}{2\left(T_{\mu c} + T_c^s\right)k_{fc}k_c^s}$ – the coefficients of the proportional and integral components of the stator current regulator; k_{fc} – the frequency converter transmission coefficient; k_c^s, T_c^s – the transmission coefficient and time constant of the current sensor.

A filter with a transfer function can be installed at the input of the circuit to reduce over-regulation: $W_c^f(p) = \dfrac{1}{T_c^s p + 1}$.

- the transmission function of the flux linkage regulator:

$$W_{lr}(p) = \frac{(T_r p + 1)k_c^s}{2\left(T_{fl}^s + 2\left(T_{\mu c} + T_c^s\right)\right)pk_{fl\,\partial n}^s k_r L_r} = k_{flp} + k_{fli}\frac{1}{p}, \tag{2.12}$$

where $k_{flp} = \dfrac{T_r k_c^s}{2T_{\mu fl}k_{fl\,\partial n}^s k_r L_r}$, $k_{fli} = \dfrac{k_c^s}{2T_{\mu fl}k_{fl\,\partial n}^s k_r L_r}$ – the coefficients of the proportional and integral components of the rotor flux linkage regulator; $T_{fl}^{cl} = T_{fl}^s + 2\left(T_{\mu c} + T_c^s\right)$; k_{fl}^s, T_{fl}^s – the transmission coefficient and the time constant of the flux linkage sensor.

A filter with a transfer function can be installed at the input of the circuit to reduce overregulation: $W_c^f(p) = \dfrac{1}{T_{fl}^s p + 1}$.

- the transmission function of the angular rotation frequency regulator:

$$W_{rfr}(p) = \frac{\left(4T_{\mu\omega}p + 1\right)Jpk_{fl}^s k_c^s}{12T_{\mu s}^2 p^2 k_s^s k_r p_n} = k_{sp} + k_{si}\frac{1}{p}, \tag{2.13}$$

where $k_{sp} = \dfrac{Jk_{fl}^s k_c^s}{3T_{\mu s}k_s^s k_r p_n}$, $k_{si} = \dfrac{Jk_{fl}^s k_c^s}{12T_{\mu s}^2 k_s^s k_r p_n}$ – the coefficients of proportional and integral components of the angular rotation frequency regulator;

$T_{\mu s} = T_s^s + 2\left(T_{\mu c} + T_c^s\right)$; k_s^s, T_s^s – the transmission coefficient and time constant of the angular rotation frequency sensor.

A filter can be installed at the circuit input to reduce overregulation: $W_c^f(p) = \dfrac{1}{T_s^s + 2\left(T_{\mu c} + T_c^s\right)p + 1}$.

2.3. A Mathematical Model of an Inductions Motor with Asymmetric Stator Windings

The modern theory of electric machines and AC drives is based on the representation of electromagnetic values by vectors. This makes it possible not only to obtain a compact form of equations, but also to build highly efficient control systems based on operations with vector variables.

Traditionally, the synthesis of new and research of known vector control algorithms are carried out on the basis of orthogonal models of IM [141, 143], in which the asymmetry of the stator windings presents significant difficulties. As the analysis of literature reveals, a complete asymmetric three-phase system can be adequately represented only in a three-dimensional space.

Thus, in [126] it was shown that when depicting a system of linearly independent winding currents $\mathbf{I}^F = \left[I_a \; I_b \; I_c\right]^T$ in the phase basis $\vec{\vec{\mathbf{F}}}$ by linearly dependent subsystems, namely the optimal magnetizing $\mathbf{I}_m^F$ and neutral $\mathbf{I}_n^F$:

$$\mathbf{I}^F = \mathbf{I}_m^F + \mathbf{I}_n^F,$$

direct conversion of the phase current vector to an orthogonal frame of reference $\mathbf{I}^G = \left[I_\alpha \; I_\beta \; I_\gamma\right]^T$ in basis $\vec{\vec{\mathbf{G}}}$ allows obtaining the transformed current vector in the form of the sum of the transformed magnetizing and transformed neutral vectors:

$$\mathbf{I}^G = \mathbf{I}_m^G + \mathbf{I}_n^G,$$

where $\mathbf{I}_m^G = \mathbf{A}_1 \mathbf{I}_m^F = \begin{bmatrix} I_\alpha & I_\beta & 0 \end{bmatrix}^T$; $\mathbf{I}_n^G = \mathbf{A}_1 \mathbf{I}_n^F = \begin{bmatrix} 0 & 0 & I_\gamma \end{bmatrix}$.

Comparing the expression of the three-dimensional radius vector of current in basis $\vec{\mathbf{G}}$ with the expression of the complex image vector of current in algebraic form:

$$\underline{I} = \frac{2}{3Q_б}\left(I_a Q_a e^{j\varphi_a} + I_b Q_b e^{j\varphi_b} + I_c Q_c e^{j\varphi_c}\right) = I_\alpha + jI_\beta,$$

where Q_a, Q_b, Q_c – the spatial winding functions of phase windings; φ_a, φ_b, φ_c – the parameters of the spatial arrangement of the magnetic axes of the phase windings in the cross-sectional plane of the machine, we can judge that the three-dimensional radius-vector completely determines the components of the complex vector by its coordinates I_α and I_β. However, the latter with its components characterizes the coordinates of the three-dimensional radius vector to the neutral component, i.e., determines only the magnetizing component of the current vector.

Using a different approach in [7] based on the concept of the resulting vector, it is shown that when representing a three-phase asymmetric system with the sum of three symmetric components (forward, inverse and zero sequence), the generalized current vector will be determined by the sum of forward and inverse components:

$$\boldsymbol{i} = \frac{2}{3}\left[\left(i_{va} + i_{za} + i_0\right) + \left(i_{vb} + i_{zb} + i_0\right)\boldsymbol{a} + \left(i_{vc} + i_{zc} + i_0\right)\boldsymbol{a}^2\right] =$$
$$= \frac{2}{3}\left[\left(i_{va} + i_{za}\right) + \left(i_{vb} + i_{zb}\right)\boldsymbol{a} + \left(i_{vc} + i_{zc}\right)\boldsymbol{a}^2 + i_0\left(1 + \boldsymbol{a} + \boldsymbol{a}^2\right)\right] = \boldsymbol{i}_v + \boldsymbol{i}_z,$$

where $\boldsymbol{a} = e^{j2\pi/3}$ – a three-phase system operator; i_{va}, i_{vb} i_{vc} – the components of the forward sequence of the stator current; i_{za}, i_{zb}, i_{zc} – stator current reverse sequence components; i_0 – the component of the zero sequence of the stator current; $\boldsymbol{i}_v$, $\boldsymbol{i}_z$ – the vectors of forward and reverse stator current sequences.

It follows from the above expressions that the generalized current vector does not contain a zero component, and it must be taken into account when analyzing any type of asymmetry.

These circumstances reflect the fact that a comprehensive description of three-dimensional problems using two-dimensional reference systems is fundamentally impossible. The use in this case of a three-dimensional frame of reference fully corresponds to the dimension of the problem.

IM mathematical model in a three-phase coordinate system is usually a system of magnetically linked windings located on the stator and rotor [144, 145]. The description of transients in IM is based on the equations of electrical equilibrium for all circuits and the equation of motion of the rotor.

The system of equilibrium equations for the stator and rotor is of the form:

$$\begin{cases} u_A = i_A R_A + \dfrac{d\Psi_A}{dt}; \\ u_B = i_B R_B + \dfrac{d\Psi_B}{dt}; \\ u_C = i_C R_C + \dfrac{d\Psi_C}{dt}; \end{cases} \begin{cases} 0 = i_a R_a + \dfrac{d\Psi_a}{dt}; \\ 0 = i_b R_b + \dfrac{d\Psi_b}{dt}; \\ 0 = i_c R_c + \dfrac{d\Psi_c}{dt}, \end{cases} \tag{2.14}$$

where u_A, u_B, u_C – stator voltages; i_A, i_B, i_C, i_a, i_b, i_c – stator and rotor currents; Ψ_A, Ψ_B, Ψ_C, Ψ_a Ψ_b, Ψ_c – stator and rotor phases flux linkages; R_A, R_B, R_C, R_a, R_b, R_c – stator and rotor phases resistances.

Flux linkage in any IM phase can be determined by the value of the intrinsic inductance of the winding and the mutual inductance with all other windings. Based on this, the flux linkages of IM phases are determined as follows:

$$\begin{cases} \Psi_A = L_A i_A + M_{AB} i_B + M_{AC} i_C + M_{Aa} i_a + M_{Ab} i_b + M_{Ac} i_c; \\ \Psi_B = L_B i_B + M_{BA} i_A + M_{BC} i_C + M_{Ba} i_a + M_{Bb} i_b + M_{Bc} i_c; \\ \Psi_C = L_C i_C + M_{CA} i_A + M_{CB} i_B + M_{Ca} i_a + M_{Cb} i_b + M_{Cc} i_c; \\ \Psi_a = L_a i_a + M_{ab} i_b + M_{ac} i_c + M_{aA} i_A + M_{aB} i_B + M_{aC} i_C; \\ \Psi_b = L_b i_b + M_{ba} i_a + M_{bc} i_c + M_{bA} i_A + M_{bB} i_B + M_{bC} i_C; \\ \Psi_c = L_c i_c + M_{ca} i_a + M_{cb} i_b + M_{cA} i_A + M_{cB} i_B + M_{cC} i_C, \end{cases} \tag{2.15}$$

where L_A, L_B, L_C, L_a, L_b, L_c – the inductances of the stator and rotor phases of the motor; M_{xy} – windings x and y mutual inductance.

The inductances of the stator and rotor phases consist of the main inductances and scattering inductances (index σ), and are determined as follows:

$$\begin{aligned} &L_A = L_{AA} + L_{A\sigma};\ L_B = L_{BB} + L_{B\sigma};\ L_C = L_{CC} + L_{C\sigma};\\ &L_a = L_{aa} + L_{a\sigma};\ L_b = L_{bb} + L_{b\sigma};\ L_c = L_{cc} + L_{c\sigma}. \end{aligned} \tag{2.16}$$

The main inductances of the stator and rotor phases under the symmetry of IM windings are the same and do not depend on the angular position of the rotor:

$$L_{AA} = L_{BB} = L_{CC} = L_{aa} = L_{bb} = L_{cc} = M,$$

where M – the maximum value of mutual inductance related to the mutual inductance between the stator and rotor by expression $M = (2/3)L_{\mu}$ [146].

When determining the mutual inductance between the stator and rotor windings, it is necessary to take into account that their spatial location changes, as a result of which its value also changes. The maximum value of mutual inductance corresponds to the coincidence of the axes of the two phases, and with the perpendicular arrangement of the axes, it is zero. Therefore, the mutual inductance between the stator and rotor windings will vary according to the harmonic law and can be calculated by expressions [147, 148]:

$$\begin{cases} M_{Aa} = M_{Bb} = M_{Cc} = M_{aA} = M_{bB} = M_{cC} = M\cos\gamma; \\ M_{Ab} = M_{Bc} = M_{Ca} = M_{bA} = M_{cB} = M_{aC} = M\cos\left(\gamma + 2\pi/3\right); \\ M_{Ac} = M_{Ba} = M_{Cb} = M_{cA} = M_{aB} = M_{bC} = M\cos\left(\gamma - 2\pi/3\right) \end{cases} \tag{2.17}$$

where γ – rotor rotation angle.

Taking into account equations 2.16–2.17 and the condition that when connecting the stator windings as a "star," the sum of phase currents is zero ($i_A + i_B + i_C = 0$), the expression for determining the flux linkage of the stator phase "A" is as follows:

$$\Psi_A = \left(L_\mu + L_{A\sigma}\right) i_A + \\ + \frac{2}{3} L_\mu \left(i_a \cos(\gamma) + i_b \cos\left(\gamma + \frac{2\pi}{3}\right) + i_c \cos\left(\gamma - \frac{2\pi}{3}\right)\right) \tag{2.18}$$

To take into account changes in IM electromagnetic parameters in the presence of asymmetry of the stator phase windings, the asymmetry coefficient ε_w can be used. It is determined as follows:

$$\varepsilon_w = \frac{w_d}{w_h}, \tag{2.19}$$

where w_d – the number of turns in the damaged phase; w_h – the number of turns in the undamaged phase.

According to [50], the resistance of the stator windings is determined as:

$$R_s = \rho l_\delta w_1 / s_n , \tag{2.20}$$

where ρ – resistivity; w_1 – the number of turns in the stator phase; l_δ – the calculated length of the magnetic circuit; s_n – conductor cross section.

When the number of turns in the stator phase changes, value $\rho l_\delta / s_n$ remains unchanged and the number of turns is determined as $w_1 \varepsilon_w = w_d$, then the resistance of the stator phase is determined as:

$$R_s = \frac{\varepsilon_w w_1 \rho l_\delta}{s} . \tag{2.21}$$

The scattering inductance of the stator windings is determined as [50]:

$$L_{s\sigma} = \frac{m_1}{2\pi f_1} w_1^2 k_{w1}^2 X_1 , \tag{2.22}$$

where m_1 – the number of the stator phases; k_{w1} – stator phase winding coefficient; X_1 – the stator winding coefficient that is determined as follows:

$$X_1 = \frac{4 f_1}{\pi} \frac{\mu_0 \tau l_\delta}{p_n k_\mu k_\delta \delta},$$

where μ_0 – magnetic permeability; τ – pole division; p_n – the number of pole pairs; k_δ – Carter's coefficient; δ – the width of the gap.

In the presence of asymmetry of the stator winding expression $\frac{m_1}{2\pi f_1} k_{w1}^2 X_1$ remains unchanged, and then one can write:

$$L_{s\sigma} = \varepsilon_w^2 w_1^2 \frac{m_1}{2\pi f_1} k_{w1}^2 X_1 . \tag{2.23}$$

The inductance of the magnetization circuit is determined as [50]:

$$L_\mu = \frac{m_1}{2\pi f_1} w_1 k_{w1} w_2 k_{w2} k_c X_1 , \tag{2.24}$$

where w_2 – the number of turns in the rotor phase; k_{w2} – the rotor phase winding coefficient; k_c – skew coefficient.

With an asymmetrical winding value $\frac{m_1}{2\pi f_1} k_{w1} w_2 k_{w2} k_c X_1$ is a constant value, and the magnetization inductance is determined as:

$$L_\mu = \varepsilon_w w_1 \frac{m_1}{2\pi f_1} k_{w1} w_2 k_{w2} k_c X_1 . \tag{2.25}$$

Thus, the presence of asymmetry of the windings of one of the phases of the IM stator can be taken into account in the expressions of electromagnetic parameters as follows:

$$r_s = \varepsilon_w R_s ; \; l_{s\sigma} = \varepsilon_w^2 L_{s\sigma} ; \; l_\mu = \varepsilon_w L_\mu , \tag{2.26}$$

where r_s, $l_{s\sigma}$, l_μ – electromagnetic parameters of the asymmetric phase of IM stator.

Thus, in case of asymmetry of phase "A," taking into account the coefficient of asymmetry ε_w (2.19), the equation of flux linkage of phase "A" of the stator (2.18) in the expanded variant takes the form:

$$\begin{aligned}\Psi_A &= \left(L_\mu \varepsilon_{wA} + L_{A\sigma}\varepsilon_{wA}^2\right) i_A + \\ &+ \tfrac{2}{3} L_\mu \varepsilon_{wA}\left(i_a \cos(\gamma) + i_b \cos\left(\gamma + \tfrac{2\pi}{3}\right) + i_c \cos\left(\gamma - \tfrac{2\pi}{3}\right)\right)\end{aligned} \tag{2.27}$$

It is necessary to use circuit modeling for the convenience of the research of IM processes at the supply from the frequency converter with a DC intermediate link [149, 150]. In this approach, the mathematical model contains two information-related systems: a circuit and a structural one. In the developed IM model the stator windings are a circuit part and are represented by *RL*-links with additional EMF, which are calculated in the structural part of the model (Figure 2.6). In the structural part, the angular rotation frequency of the shaft and the electromagnetic torque of the motor are also calculated.

The stator EMF of the motor is calculated by substituting the expressions to determine the phase flux linkages (2.27) into the equilibrium equation (2.14).

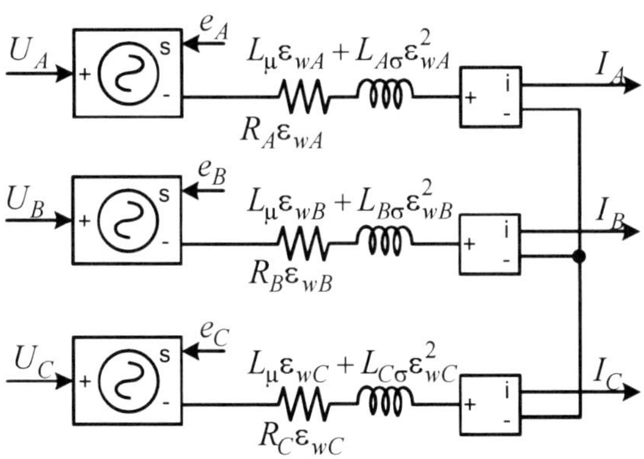

Figure 2.6. A circuit model of IM stator.

Then, taking into account the coefficient of asymmetry of the windings, we obtain:

$$u_A = i_A R_A \varepsilon_{wA} + \left(L_\mu \varepsilon_{wA} + L_{A\sigma} \varepsilon_{wA}^2\right)\frac{di_A}{dt} +$$
$$+ \frac{2}{3} L_\mu \varepsilon_{wA}\left(\frac{di_a}{dt}\cos(\gamma) + \frac{di_b}{dt}\cos\left(\gamma + \frac{2\pi}{3}\right) + \frac{di_c}{dt}\cos\left(\gamma - \frac{2\pi}{3}\right)\right) -$$
$$- \frac{2}{3} L_\mu \varepsilon_{wA}\omega_e\left(i_a \sin(\gamma) + i_b \sin\left(\gamma + \frac{2\pi}{3}\right) + i_c \sin\left(\gamma - \frac{2\pi}{3}\right)\right),$$

where ω_e – electric angular frequency of rotor rotation.

The same equation can be written as follows:

$$u_A = i_A R_A \varepsilon_{wA} + \left(L_\mu \varepsilon_{wA} + L_{A\sigma} \varepsilon_{wA}^2\right)\frac{di_A}{dt} + e_A \tag{2.28}$$

where e_A – the electromotive force of phase "A" of the stator.

The equilibrium equations for all IM phases, the calculation of stator EMF of which is based on the equations, are recorded accordingly:

$$\left.\begin{aligned}
e_A &= \frac{2}{3} L_\mu \varepsilon_{wA}\left(\begin{aligned}&\frac{di_a}{dt}\cos(\gamma) + \frac{di_b}{dt}\cos\left(\gamma + \frac{2\pi}{3}\right) + \frac{di_c}{dt}\cos\left(\gamma - \frac{2\pi}{3}\right) - \\ &-\omega_e\left(i_a \sin(\gamma) + i_b \sin\left(\gamma + \frac{2\pi}{3}\right) + i_c \sin\left(\gamma - \frac{2\pi}{3}\right)\right)\end{aligned}\right); \\
e_B &= \frac{2}{3} L_\mu \varepsilon_{wA}\left(\begin{aligned}&\frac{di_a}{dt}\cos\left(\gamma - \frac{2\pi}{3}\right) + \frac{di_b}{dt}\cos(\gamma) + \frac{di_c}{dt}\cos\left(\gamma + \frac{2\pi}{3}\right) - \\ &-\omega_e\left(i_a \sin\left(\gamma - \frac{2\pi}{3}\right) + i_b \sin(\gamma) + i_c \sin\left(\gamma + \frac{2\pi}{3}\right)\right)\end{aligned}\right); \\
e_C &= \frac{2}{3} L_\mu \varepsilon_{wA}\left(\begin{aligned}&\frac{di_a}{dt}\cos\left(\gamma + \frac{2\pi}{3}\right) + \frac{di_b}{dt}\cos\left(\gamma - \frac{2\pi}{3}\right) + \frac{di_c}{dt}\cos(\gamma) - \\ &-\omega_e\left(i_a \sin\left(\gamma + \frac{2\pi}{3}\right) + i_b \sin\left(\gamma - \frac{2\pi}{3}\right) + i_c \sin(\gamma)\right)\end{aligned}\right).
\end{aligned}\right\}$$

Solving equation (2.28) with respect to the current we obtain:

$$\frac{di_A}{dt} = \left(u_A - i_A R_A \varepsilon_{wA} - e_A\right)\frac{1}{L_\mu \varepsilon_{wA} + L_{A\sigma} \varepsilon_{wA}^2} \tag{2.29}$$

Thus, the windings of IM stator when connected in the form of a star (Figure 2.6) contain series-connected active resistors, inductors and controlled sources of

EMF. Thus, a system of equations of electrical equilibrium of the stator windings is implemented by means of circuit modeling.

The currents in IM rotor phases can be calculated by expressions:

$$\left.\begin{aligned}
i_a &= \frac{1}{L_{a\sigma}}\left(\begin{aligned}&\int\left(-i_a R_a\right)dt - \\ &- L_\mu\left(\tfrac{3}{2} i_a + i_A\cos\gamma + i_C\cos\left(\gamma + \tfrac{2\pi}{3}\right) + i_B\cos\left(\gamma - \tfrac{2\pi}{3}\right)\right)\end{aligned}\right);\\
i_b &= \frac{1}{L_{b\sigma}}\left(\begin{aligned}&\int\left(-i_b R_b\right)dt - \\ &- L_\mu\left(\tfrac{3}{2} i_b + i_B\cos\gamma + i_A\cos\left(\gamma + \tfrac{2\pi}{3}\right) + i_C\cos\left(\gamma - \tfrac{2\pi}{3}\right)\right)\end{aligned}\right);\\
i_c &= \frac{1}{L_{c\sigma}}\left(\begin{aligned}&\int\left(-i_c R_c\right)dt - \\ &- L_\mu\left(\tfrac{3}{2} i_c + i_C\cos\gamma + i_B\cos\left(\gamma + \tfrac{2\pi}{3}\right) + i_A\cos\left(\gamma - \tfrac{2\pi}{3}\right)\right)\end{aligned}\right).
\end{aligned}\right\};$$

When using structural modeling, it is more expedient to implement IM model using a matrix representation [151]. Then the matrix of instantaneous values of mutual inductances between the stator and rotor windings M_c is of the form:

$$M_c = \begin{bmatrix} \cos\gamma & \cos\left(\gamma + \tfrac{2\pi}{3}\right) & \cos\left(\gamma - \tfrac{2\pi}{3}\right) \\ \cos\left(\gamma - \tfrac{2\pi}{3}\right) & \cos\gamma & \cos\left(\gamma + \tfrac{2\pi}{3}\right) \\ \cos\left(\gamma + \tfrac{2\pi}{3}\right) & \cos\left(\gamma - \tfrac{2\pi}{3}\right) & \cos\gamma \end{bmatrix}.$$

The matrix of instantaneous values of mutual inductances (M_s), orthogonal to the coefficients of matrix (M_c) is of the form:

$$M_s = \begin{bmatrix} \sin\gamma & \sin\left(\gamma + \tfrac{2\pi}{3}\right) & \sin\left(\gamma - \tfrac{2\pi}{3}\right) \\ \sin\left(\gamma - \tfrac{2\pi}{3}\right) & \sin\varphi_e & \sin\left(\gamma + \tfrac{2\pi}{3}\right) \\ \sin\left(\gamma + \tfrac{2\pi}{3}\right) & \sin\left(\gamma - \tfrac{2\pi}{3}\right) & \sin\gamma \end{bmatrix}.$$

Taking into account the matrix operations, the models for calculating the rotor current, the flux coupling and the stator EMF are given in Figure 2.7.

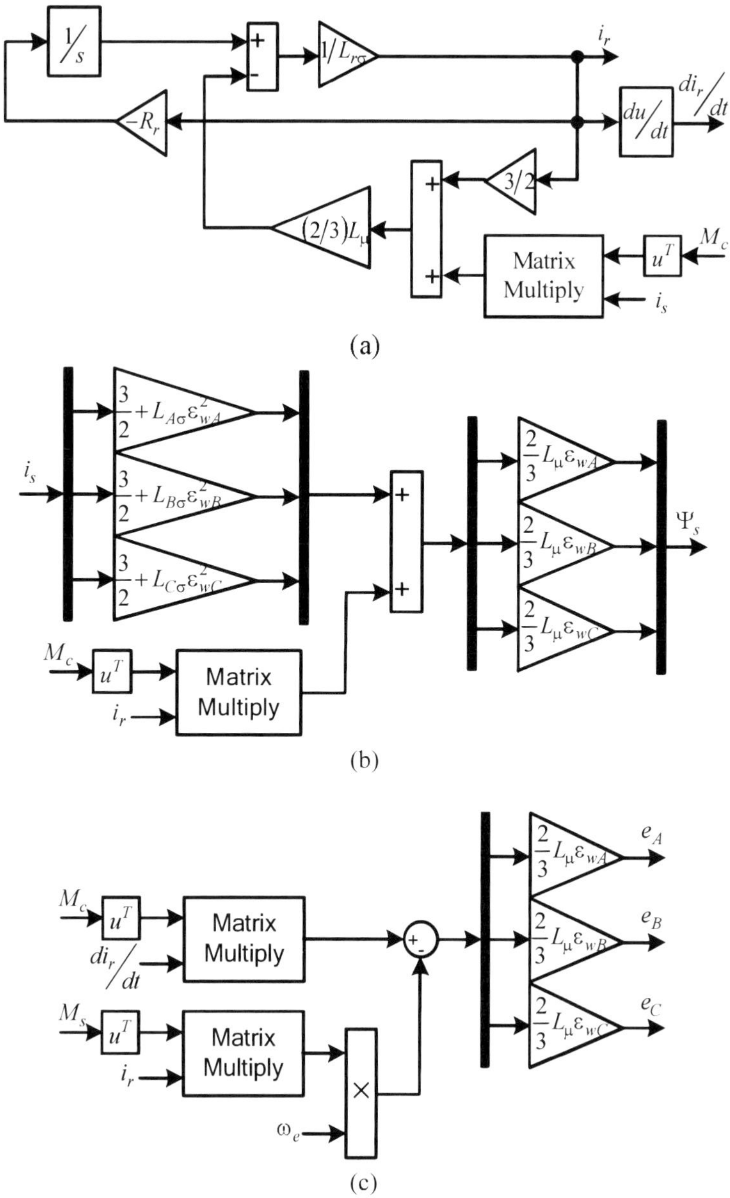

Figure 2.7. The subsystems for calculating the rotor current (a) the stator flux linkage (b) and the stator EMF (c).

Modeling of the mechanical part is carried out on the basis of the equation for the electromagnetic moment and the equation of the rotor motion (2.30), the model of which is given in Figure 2.8.

$$\begin{cases} T_e = \frac{p_n}{\sqrt{3}}\left[(\Psi_C - \Psi_B)i_A + (\Psi_A - \Psi_C)i_B + (\Psi_B - \Psi_A)i_C\right]; \\ \frac{d\omega_e}{dt} = \frac{p_n}{J}(T_e - T_l), \end{cases} \tag{2.30}$$

where ω_e – the electric angular frequency of rotor rotation; p_n – the number of IM poles pairs; J – rotor inertia moment; T_e – IM electromagnetic torque; T_l – load moment.

The model of the subsystem for calculating the instantaneous values of the mutual inductances between the stator and rotor windings (M_c) and the mutual inductances coefficients orthogonal to them (M_s) are shown in Figure 2.9.

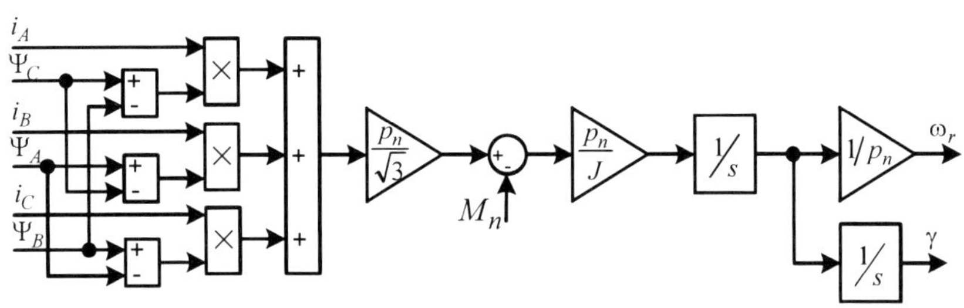

Figure 2.8. The model of IM mechanical part.

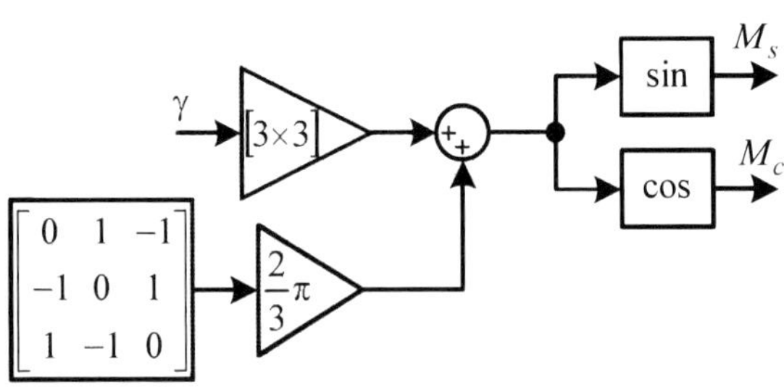

Figure 2.9. A subsystem for calculating the instantaneous values of the mutual inductances.

Thus, the system of differential equations of IM in the three-phase coordinate system can be used to calculate the processes in motors with asymmetric stator windings when supplied by FC.

The verification of the reliability of the presented mathematical model of IM with asymmetry of stator windings was performed by comparing the experimental and model signals of current and electromagnetic torque of IM series AIR80V4U2 with the parameters: $P_n = 1.5$ kW, $I_n = 3.6$ A, $n_n = 1395$ rev/min, $\eta = 0.77$, $\cos\varphi = 0.81$, $R_s = 5$ Ohm, $R_r = 3.69$ Ohm, $X_s = 4.35$ Ohm, $X_r = 4$ Ohm, $X_\mu = 58.6$ Ohm.

The simulation of asymmetry in phase "A" of the stator of the researched IM is carried out with the help of solders, the connection scheme of which is shown in Figure 2.10. It allowed researching the modes of IM operation at the following values of asymmetry of the electromagnetic parameters of the motor stator: 1 – 1.36% ($\varepsilon_{wA} = 0.9864$); 2 – 2.52% ($\varepsilon_{wA} = 0.9748$); 3 – 14% ($\varepsilon_{wA} = 0.86$).

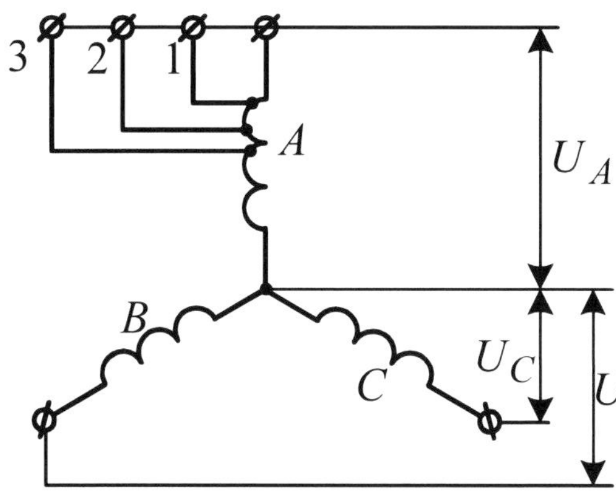

Figure 2.10. The circuit of soldering connection on IM stator.

Measurements and recording of current and voltage signals were performed using the developed measuring module [152]. The experimental signals of voltages and currents of the phases of IM stator are shown in Figure 2.11.

To bring the modeling results closer to the real operating conditions of the experimental IM, the calculations in the mathematical model were performed using experimentally measured instantaneous values of the stator phase voltages. (Figure 2.11, a).

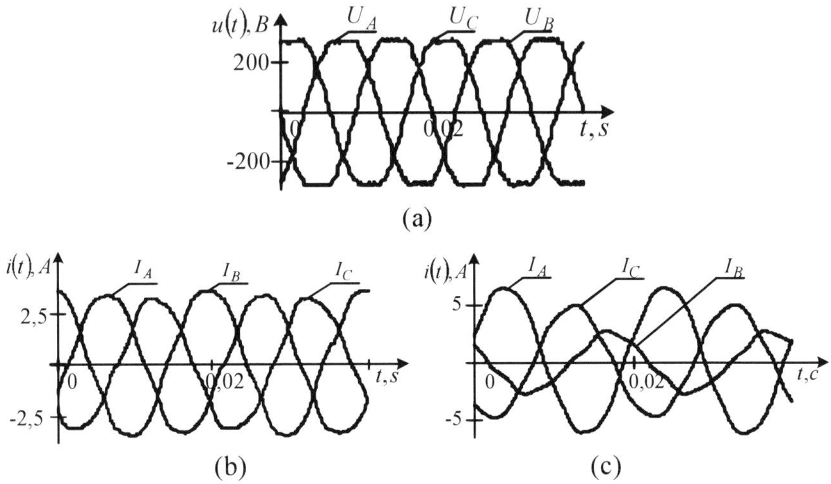

Figure 2.11. The experimental signals of voltages (a) and currents of IM stator in a symmetrical machine (b) and at 14% asymmetry of the stator phase (c).

The comparison of the signals of the stator phase currents and the electromagnetic torque of the researched IM is shown in Figures 2.12–2.17.

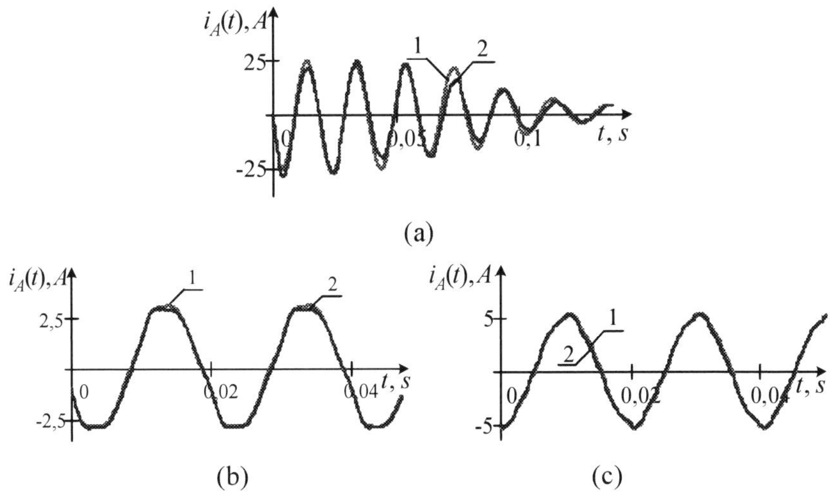

Figure 2.12. Experimental (1) and model (2) phase A current signals of an induction motor with symmetrical stator windings: a) at the start; b) in the idle mode; c) at the rated load.

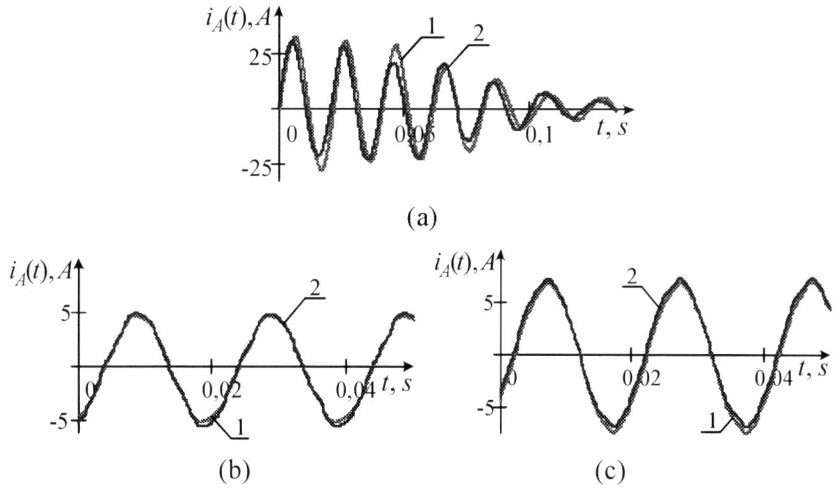

Figure 2.13. Experimental (1) and model (2) current signals of phase A of an induction motor at 14% stator phase asymmetry: a) at the start; b) in the idle mode; c) at the rated load.

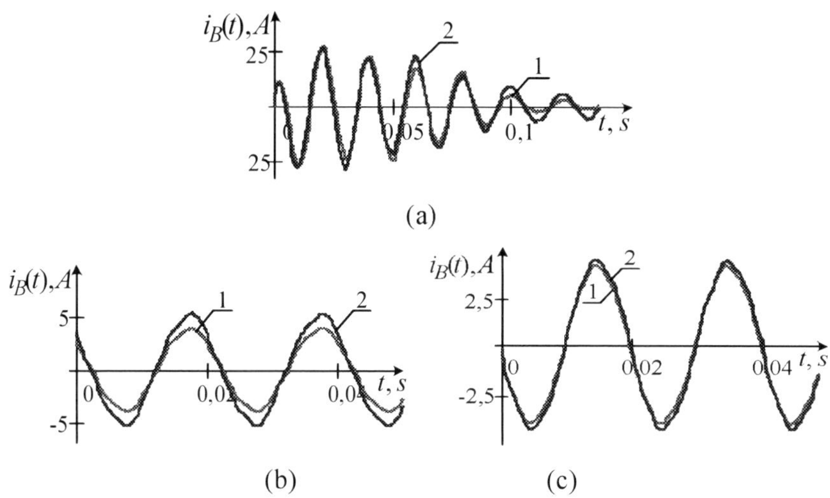

Figure 2.14. Experimental (1) and model (2) phase B current signals of an induction motor at 14% stator phase asymmetry: a) at the start; b) in the idle mode; c) at the rated load.

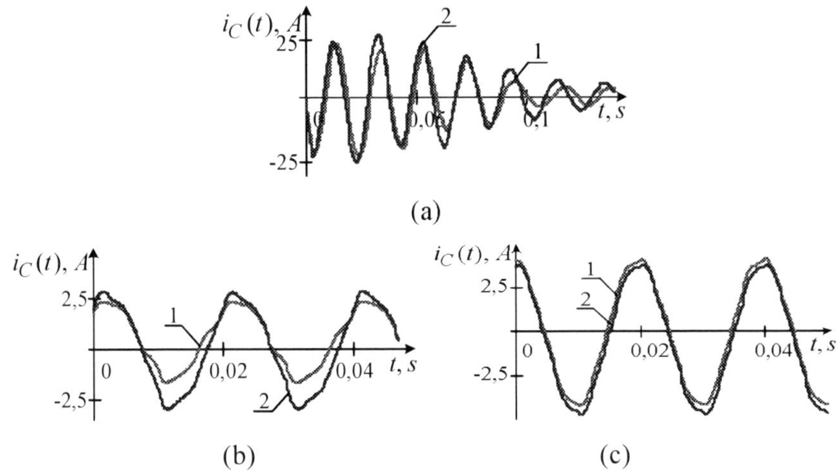

Figure 2.15. Experimental (1) and model (2) phase B current signals of an induction motor at 14% stator phase asymmetry: a) at the start; b) in the idle mode; c) at the rated load.

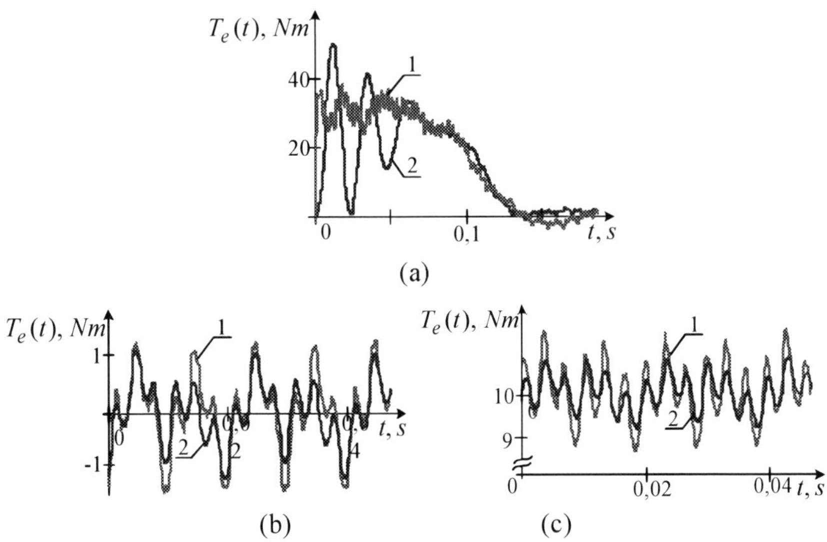

Figure 2.16. Calculated from the experimental current and voltage signals (1) and model (2) signals of the electromagnetic torque of an induction motor with symmetrical windings: a) at the start; b) in the idle mode; c) at the rated load.

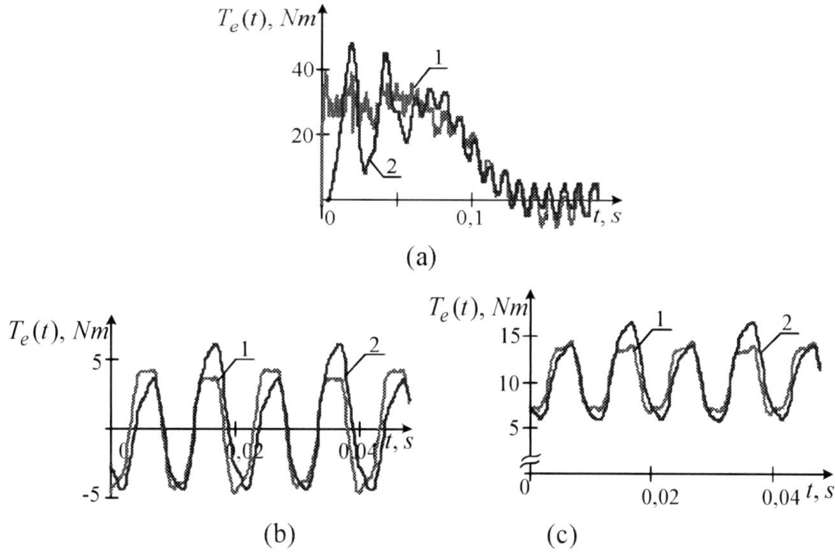

Figure 2.17. Calculated from the experimental current and voltage signals (1) and model (2) signals of the electromagnetic torque at 14% asymmetry of the stator phase: a) at the start; b) in the idle mode; c) at the rated load.

The coefficient of determination was used to assess the deviations of the model signals from the real ones:

$$R^2 = 1 - \frac{\sum_{i=0}^{N-1} \left(Y_i - Y_r(i\Delta t)\right)^2}{\sum_{i=0}^{N-1} \left(Y_i - Y_s\right)^2},$$

where Y_i – the instantaneous value of the real signal; $Y_r(i\Delta t)$ – the function that describes the instantaneous value of the calculated signal

(Δt – the sampling interval of the real signal); Y_s – the average value of the real signal.

Table 2.1 shows the deviation of the instantaneous values of the calculated signals from the experimental ones in the idle mode, when operating with the rated load and when starting the IM.

Table 2.1. The deviation of the calculated instantaneous signal values from the experimental ones

Operating conditions	R^2			
	I_A	I_B	I_C	T_e
	Symmetrical motor			
Idle mode	0.996	0.997	0.998	0.777
Load	0.995	0.997	0.995	0.701
Start	0.946	0.942	0.952	0.531
	Stator asymmetry of 1.36%			
Idle mode	0.998	0.995	0.993	0.88
Load	0.993	0.997	0.997	0.855
Start	0.942	0.952	0.961	0.561
	Stator asymmetry of 2.52%			
Idle mode	0.997	0.994	0.986	0.805
Load	0.99	0.997	0.998	0.775
Start	0.872	0.935	0.921	0.541
	Stator asymmetry of 14%			
Idle mode	0.967	0.934	0.952	0.699
Load	0.989	0.992	0.979	0.677
Start	0.821	0.911	0.805	0.642

The results of the research revealed that at 14% of asymmetry of the stator windings one of IM phases operates in the generator mode and gives both active and reactive power to the mains.

The results of the comparison of the instantaneous values of the calculated signals with experimental ones showed that the presented mathematical model of induction motor in three-phase coordinate system makes it possible to adequately reproduce electromagnetic and electromechanical processes occurring in real motors with stator windings asymmetry.

2.4. The Assessment of the Operation Modes of the Variable-Frequency Electric Drive with a Vector Control

The research of the operation modes of a variable-frequency ED with a vector control was performed on mathematical models in which the frequency converter was represented by a model containing a three-phase bridge uncontrolled rectifier, an *LC*-filter in the DC link and a three-phase voltage inverter on IGBT with reverse diodes; IM represented by a mathematical model in a three-phase coordinate system, in which the asymmetry of the stator windings takes into account the change in active resistance, scattering inductance and main mutual induction [50]; the control system oriented by the rotor flux linkage vector and synthesized in an orthogonal coordinate system. For the control problems of IM in a three-phase coordinate system, the theory of direct and inverse coordinate and phase transformations was used [136].

A modified PWM method was used to generate the control signal pulses of the autonomous voltage inverter [153, 154], the essence of which is to apply a non-sinusoidal law to modulate the pulse duration of the inverter output voltage. The main task of this method of generating signals of power transistors control is to increase the ratio of the fundamental harmonics of the output voltage of the inverter to the voltage of the power supply. Among the laws of modulation that solve this problem, one can distinguish modulation by rectangular and trapezoidal laws. However, when using them, the harmonic composition of the phase and linear output voltage of the inverter deteriorates: in the low-frequency region of the spectrum there are distortion harmonics, first of all, the 5 th, 7 th, 11 th and 13 th harmonics of the output frequency.

The method of introducing the third harmonic in the reference signals is the most effective and at the same time relatively simple one in the implementation of the methods of pre-modulation (Figure 2.18).

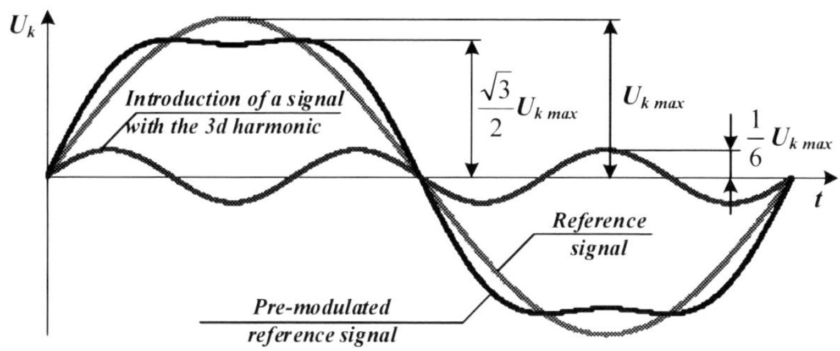

Figure 2.18. The pre-modulation of the reference signal.

When implementing this type of PWM in the control signal supplied to the inputs of the three-phase PWM modulator of each phase, an additional signal is added. It contains the third harmonic of the fundamental frequency in the following proportions:

$$u_{yA} = \frac{k_m u_m}{\cos(\pi/6)}\left[\sin(\omega_m t) + \left(1 - \cos\left(\frac{\pi}{6}\right)\right)\sin(3\omega_m t)\right];$$
$$u_{yB} = \frac{k_m u_m}{\cos(\pi/6)}\left[\sin\left(\omega_m t - \frac{2}{3}\pi\right) + \left(1 - \cos\left(\frac{\pi}{6}\right)\right)\sin(3\omega_m t)\right]; \qquad (2.31)$$
$$u_{yC} = \frac{k_m u_m}{\cos(\pi/6)}\left[\sin\left(\omega_m t - \frac{4}{3}\pi\right) + \left(1 - \cos\left(\frac{\pi}{6}\right)\right)\sin(3\omega_m t)\right],$$

where k_m – modulation coefficient; u_m – the maximum allowable amplitude of the control signal; ω_m – the output frequency of FC voltage.

The use of pre-modulation of the third harmonic not only increases the maximum allowable ratio of the amplitude of the first harmonic to the voltage of power supplies by 15.47%, but also leads to a significant reduction in higher harmonics in current and voltage signals.

The block diagram of the developed mathematical model of variable-frequency ED with vector control is shown in Figure 2.19.

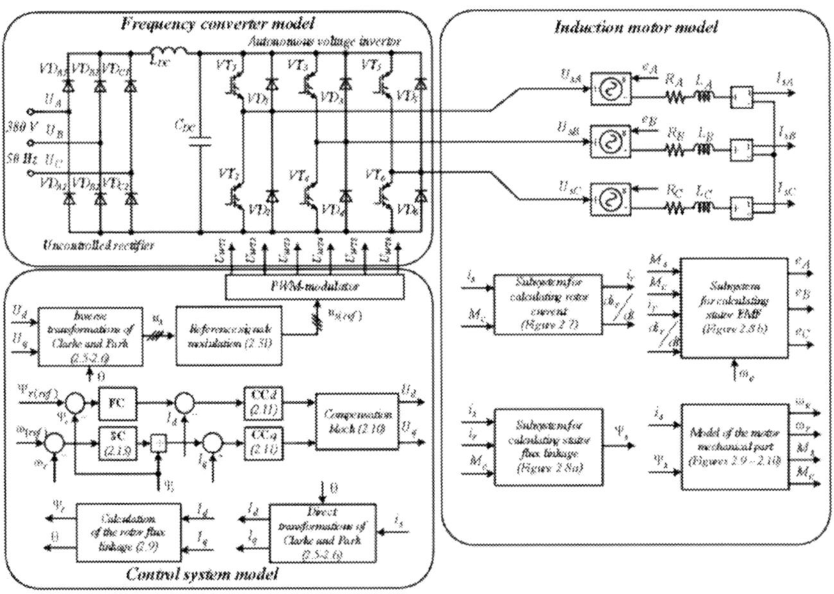

Figure 2.19. Block diagram of the model of variable-frequency ED with vector control.

The modes of operation of variable-frequency ED were researched on the developed mathematical model for the 4A112M4 series motor with the following published data: $P_n = 5.5$ kW; $n_n = 1446$ rev/min; $\cos\varphi = 0.85$; $\eta = 0.855$; $R_s = 1.228$ Ohm; $R_r = 0.787$ Ohm; $L_s = 4.76$ mHn; $L_r = 7.94$ mHn; $L_\mu = 0.171$ Hn. Diodes VS-15ETH06PBF with the following parameters were selected as the power keys of the uncontrolled rectifier in FC mathematical model: direct current $I_{vd} = 15$ A, voltage in the open state $U_{vd} = 1.3$ V; IRG4PC50UD transistors with the following parameters were used as power switches of the autonomous voltage inverter: collector current $I_c = 27$ A; collector-emitter voltage $U_{ce} = 1.65$ V; direct current of the reverse diode $I_{vd} = 25$ A, voltage in the open state of the reverse diode $U_{vd} = 1.2$ V.

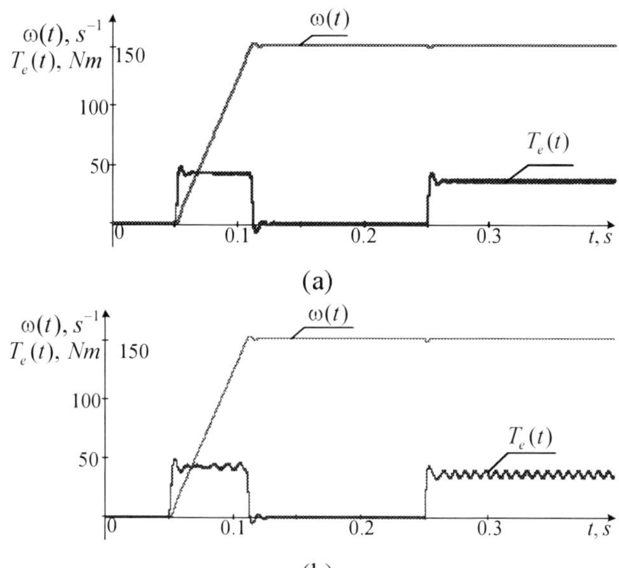

Figure 2.20. Transients in the angular frequency of rotation and the torque of symmetric (a) and asymmetric (b) IM in the ED with vector control.

Graphs of transients according to the angular frequency of rotation and the electromagnetic moment of symmetric and asymmetric IM in the ED with vector control during the start without load and subsequent load growth are shown in Figure 2.20 (the results of the research of asymmetric modes of ED operation are given for the case of asymmetry in phase A equal to 10%,

$\varepsilon_{wA} = 0.9$). In modeling, 4 kHz modulation frequency of the power keys of the autonomous voltage inverter was chosen.

Usually in the synthesis of new algorithms of IM vector control and research of the known ones the assessment of the quality of transient processes, static and dynamic error of responding the set and disturbing influences is carried out [155, 156], but the processes of energy conversion in these systems are not considered.

Taking into account the trend of expanding the scope of the use of vector control systems, which is characterized by an increase in the power of the installed converters, such ED systems energy is of significant interest. The following energy parameters were used to assess the quality of IM control [157]: losses (ΔP_{Cu1}) in the stator copper for each phase of the motor separately, losses (ΔP_{ss}

the variable component of IM electromagnetic torque ($\tilde{T}_e$) and the relative values of the active power component consumed from the network ($\tilde{p}$) and active power consumed from FC ($\tilde{p}_{fc}$).

2.4.1. The Assessment of Losses in the Power Part of the Variable-Frequency ED

Since in variable-frequency ED systems the voltage is formed by the PWM method, the effect of higher voltage harmonics on energy losses in IM stator windings was researched. Thus, the voltage signals of phase A and the current of the stator phases of the researched motor as part of ED with vector control are shown in Figure 2.21.

The following methods were considered to calculate the losses in the copper of IM stator windings:

- the calculation of instantaneous values of phase current [158] (method 1):

$$\begin{cases} \Delta p_{Cu1}(t) = i_A(t)^2 R_A + i_B(t)^2 R_B + i_C(t)^2 R_C; \\ \Delta P_{Cu1} = \dfrac{1}{T}\displaystyle\int_0^T \Delta p_{Cu1}, \end{cases} \tag{2.32}$$

where $i_A(t)$, $i_B(t)$, $i_C(t)$ – current instantaneous values.

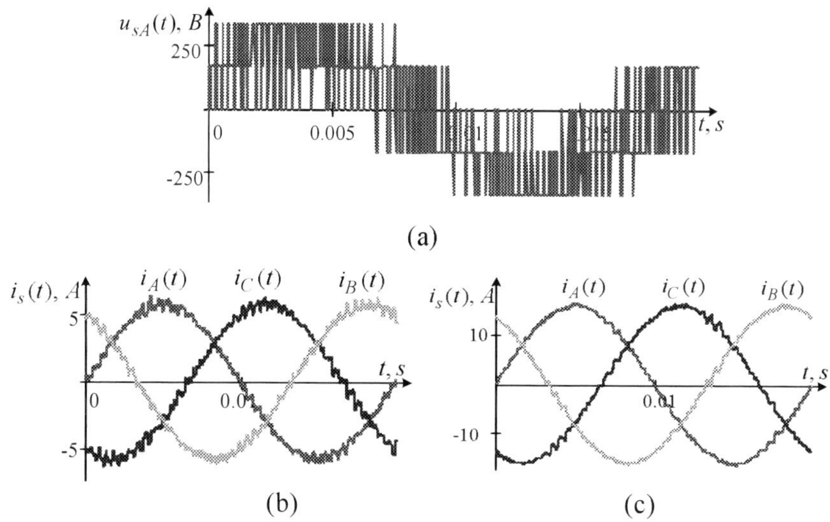

Figure 2.21. IM phase supply voltage (a) and stator current in idle mode (b) and at rated load (c).

- the calculation based on the values of the amplitude of the fundamental harmonic of the stator current and the value of the amplitude of the harmonic current and stator resistance at the PWM frequency [159] (method 2):

$$\begin{cases} \Delta P_{Cu1} = I_1^2 R_s + I_{pwm}^2 R_{pwm}; \\ R_{pwm} = R_s \sqrt{\dfrac{f_{pwm}}{f_1}}, \end{cases} \tag{2.33}$$

where I_1 – the amplitude of the fundamental harmonic of the stator current, I_{pwm} – current harmonic amplitude with PWM frequency, f_{pwm} – PWM frequency.

- the calculation based on the values of the amplitude of the fundamental harmonic of the stator current and the equivalent value of the stator resistance [160, 161] (method 3):

$$\begin{cases} \Delta P_{Cu1} = I_1^2 R_s'; \\ R_s' = R_s\left[1+\delta\left(\dfrac{\Delta I_P}{I_1}\right)^2\right], \end{cases} \tag{2.34}$$

where δ – reduction factor; ΔI_P – stator current pulsations.

When calculating the losses in IM stator copper by the latter method according to [161] the following was assumed: reduction coefficient $\delta = 1/12$ (for motors of the power up to 15 kW); the amplitude of the current fundamental harmonic for phase A in idle mode $I_1 = 5.815$ A, stator current pulsations $\Delta I_P = 0.667$ A (the relation of which to the amplitude of current fundamental harmonic will be equal to $\Delta I_{\Pi} = 0.115$); the amplitude of current fundamental harmonic for phase A when working with the rated load $I_1 = 16.076$ A, stator current pulsations $\Delta I_P = 0.565$ A (the relation of which to the amplitude of current fundamental harmonic will be equal to $\Delta I_P = 0.035$).

Comparison of the values of the symmetrical IM stator copper losses calculated by the considered methods is given in Table 2.2.

Table 2.2. Stator copper loss in symmetrical IM

Calculation methods		Losses, W			
		ΔP_{Cu1_A}	ΔP_{Cu1_B}	ΔP_{Cu1_C}	ΔP_{Cu1_Σ}
Idle mode					
Method 1 (2.32)		20.49	20.835	20.759	62.084
Method 2 (2.33)	$R_s I_1^2$	20.416	20.759	20.693	61.868
	$I_{pwm}^2 R_{pwm}$	$14.5\cdot10^{-3}$	$3.77\cdot10^{-3}$	$7.26\cdot10^{-3}$	$25.5\cdot10^{-3}$
Method 3 (2.34)	$R_s I_1^2$	20.416	20.759	20.693	61.868
	$R_s I_1^2 \delta(\Delta I_{\Pi}/I_1)^2$	$1.35\cdot10^{-3}$	$1.35\cdot10^{-3}$	$1.43\cdot10^{-3}$	$4.13\cdot10^{-3}$
Rated load					
Method 1 (2.32)		158.835	157.305	158.029	474.169
Method 2 (2.33)	$R_s I_1^2$	158.66	157.185	157.93	473.775
	$I_{pwm}^2 R_{pwm}$	$0.52\cdot10^{-3}$	$0.4\cdot10^{-3}$	$0.69\cdot10^{-3}$	$1.24\cdot10^{-3}$
Method 3 (2.34)	$R_s I_1^2$	158.66	157.185	157.93	473.775
	$R_s I_1^2 \delta(\Delta I_{\Pi}/I_1)^2$	$0.13\cdot10^{-3}$	$0.12\cdot10^{-3}$	$0.13\cdot10^{-3}$	$0.38\cdot10^{-3}$

The comparison of the obtained results revealed that the errors of calculation of stator copper losses by instantaneous current signals on average do not exceed 0.3% in comparison with method 2, and 0.35% in comparison with method 3, in idle mode, and 0.1% in comparison with methods 2 and 3 when working with rated load.

The comparison of IM stator copper losses at asymmetry in phase A, equal to 10%, according to the considered methods is given in Table 2.3.

The results of the calculation of induction motor losses showed that the errors in determining the losses in the stator copper by instantaneous current signals do not exceed 0.31% on average compared with method 2, and 0.35% compared with method 3, in idle mode, and 0.26% compared with methods 2 and 3 when operating with rated load.

Thus, when supplying IM stator windings from the frequency converter, the increase in stator copper losses from higher harmonics of pulse-width modulated voltage is insignificant compared with the increase in losses caused by the asymmetry of stator windings, so the losses from higher harmonics can be neglected during the calculation.

Table 2.3. Induction IM stator copper losses

Calculation methods		**Losses, W**			
		ΔP_{Cu1_A}	ΔP_{Cu1_B}	ΔP_{Cu1_C}	ΔP_{Cu1_Σ}
Idle mode					
Method 1 (2.31)		19.238	22.061	20.826	62.126
Method 2 (2.32)	$R_s I_1^2$	19.145	21.999	20.76	61.904
	$I_{pwm}^2 R_{pwm}$	$18.4 \cdot 10^{-3}$	$5 \cdot 10^{-3}$	$7.43 \cdot 10^{-3}$	$30.9 \cdot 10^{-3}$
Method 3 (2.33)	$R_s I_1^2$	19.145	21.998	20.76	61.904
	$R_s I_1^2 \delta(\Delta I_\Pi / I_1)^2$	$1.35 \cdot 10^{-3}$	$1.35 \cdot 10^{-3}$	$1.43 \cdot 10^{-3}$	$4.13 \cdot 10^{-3}$
Rated load					
Method 1 (2.31)		158.587	180.56	166.716	505.862
Method 2 (2.32)	$R_s I_1^2$	158.013	180.279	166.261	504.55
	$I_{pwm}^2 R_{pwm}$	$2.5 \cdot 10^{-3}$	$9.6 \cdot 10^{-5}$	$3.33 \cdot 10^{-3}$	$5.92 \cdot 10^{-3}$
Method 3 (2.33)	$R_s I_1^2$	158.013	180.278	166.261	504.552
	$R_s I_1^2 \delta(\Delta I_\Pi / I_1)^2$	$0.13 \cdot 10^{-3}$	$0.118 \cdot 10^{-3}$	$0.133 \cdot 10^{-3}$	$0.38 \cdot 10^{-3}$

During the operation of variable-frequency ED, electric power (energy losses) is emitted in FC power semiconductor switches. It is dissipated in the form of heat [162, 163]. These losses can be divided into losses in transients of on and off (dynamic losses), losses in the on or off state (static losses) and losses in the input control circuit.

Static losses are due to the fact that in the on state the current is determined by the load parameters, and the forward voltage of the transistor is minimal. In the off state, the current of the transistor is minimal and is determined by its internal resistance, and the voltage is set by the power supply. Thus, this component of losses depends on the selected type of the semiconductor key and the load current.

The losses in the input control circuit are related to the amount of charge accumulated in the input capacitance of the gate of the transistor key, and the amount of control voltage required to switch the key. Since the power IGBT control current is relatively small, the value of these losses depends on the selected type of transistor and does not significantly affect the overall power loss [164].

Dynamic key losses are due to the fact that in the transient process the transistor is for some time in a state of high voltage and significant direct current. The switching speed of transistors with an isolated gate is regulated by switching on in series between the output node of the driver and the input circuit of the key of the limiting (gate) resistor [165]. In this case, the gate resistor is selected in such a way that the amount of dynamic losses does not significantly affect the total energy loss in the keys.

During the research of the modes of operation of the variable-frequency ED with induction IM, the static energy losses in the power switches of the uncontrolled rectifier and autonomous voltage inverter were studied. These energy losses can be determined via the instantaneous values of current through semiconductor switches [166]:

$$\Delta P_{ss} = \frac{1}{T}\int_{0}^{T} i_{ss}(t)^2 r_{ss}, \tag{2.35}$$

where $i_{ss}(t)$ – the instantaneous value of current through a semiconductor switch; r_{ss} – the resistance of the semiconductor key in the open state.

The results of the calculation of energy losses in FC power semiconductor switches when working with a symmetric IM are presented in Table 2.4, and the results of the operation with an asymmetric motor are given in Table 2.5.

Table 2.4. The losses in semiconductor switches of FC with symmetric IM

Operating conditions	**Losses, W**					
Power diodes of an uncontrolled rectifier						
	VD_{A1}	VD_{A2}	VD_{B1}	VD_{B2}	VD_{C1}	VD_{C2}
Idle mode	0.018	0.018	0.018	0.018	0.018	0.018
Rated load	4.203	4.193	4.188	4.2	4.197	4.196
Autonomous inverter transistor switches						
	VT_1	VT_2	VT_3	VT_4	VT_5	VT_6
Idle mode	0.281	0.259	0.258	0.285	0.27	0.263
Rated load	3.648	3.44	3.627	3.629	3.439	3.639
Reverse diodes of transistor switches of an autonomous inverter						
	VD_1	VD_2	VD_3	VD_4	VD_5	VD_6
Idle mode	0.187	0.192	0.206	0.187	0.185	0.2
Rated load	0.299	0.319	0.328	0.315	0.326	0.323

Table 2.5. The losses in semiconductor switches of FC with asymmetric IM

Operating conditions	**Losses, W**					
Power diodes of an uncontrolled rectifier						
	VD_{A1}	VD_{A2}	VD_{B1}	VD_{B2}	VD_{C1}	VD_{C2}
Idle mode	0.018	0.018	0.017	0.017	0.019	0.019
Rated load	3.459	3.571	4.492	4.407	5.295	5.267
Autonomous inverter transistor switches						
	VT_1	VT_2	VT_3	VT_4	VT_5	VT_6
Idle mode	0.264	0.244	0.272	0.3	0.27	0.265
Rated load	3.544	3.426	4.117	4.086	3.589	3.831
Reverse diodes of transistor switches of an autonomous inverter						
	VD_1	VD_2	VD_3	VD_4	VD_5	VD_6
Idle mode	0.175	0.182	0.222	0.201	0.183	0.199
Rated load	0.369	0.397	0.386	0.389	0.341	0.336

The results of the research demonstrate that during the operation of a FC with an induction motor, the losses in the power diodes of the uncontrolled rectifier of the most overloaded phase increase by 25%, in the power transistors of the autonomous inverter they increase by 13% and in reverse diodes – by 23%. Taking into account the fact that the energy of power key losses is released in the semiconductor crystal and dissipated as heat, long-term FC operation in this mode can lead to significant overheating of individual keys, and subsequently to their complete failure.

2.4.2. The Assessment of Variable Components of Power Consumption and Electromagnetic Torque of IM

The research reveals that the presence of asymmetry in IM stator windings leads to uneven distribution of losses in the copper of the stator by motor phases and to the deterioration of control due to pulsations of electromagnetic torque and the appearance of a variable component of power consumption. The increase of IM vibrations can be indirectly judged by the pulsations of the electromagnetic torque.

The analysis of variable power components was performed for active power consumed from the network and consumed from the FC. Their values can be calculated on the basis of expressions:

$$\tilde{p}(t) = p(t) - \frac{1}{T}\int_0^T p(t)dt; \quad \tilde{p}_{fc}(t) = p_{fc}(t) - \frac{1}{T}\int_0^T p_{fc}(t)dt, \qquad (2.36)$$

where $p(t)$, $p_{fc}(t)$ – active power consumed from the network and from the frequency converter ($p(t) = u_A(t)i_A(t) + u_B(t)i_B(t) + + u_C(t)i_C(t)$; $p_{fc}(t) = u_{sA}(t)i_{sA}(t) + u_{sB}(t)i_{sB}(t) + u_{sC}(t)i_{sC}(t)$); T – the period of the variable power component signal.

The signals of the active power consumed from the network and from FC by a symmetric motor are given in Figure 2.22.

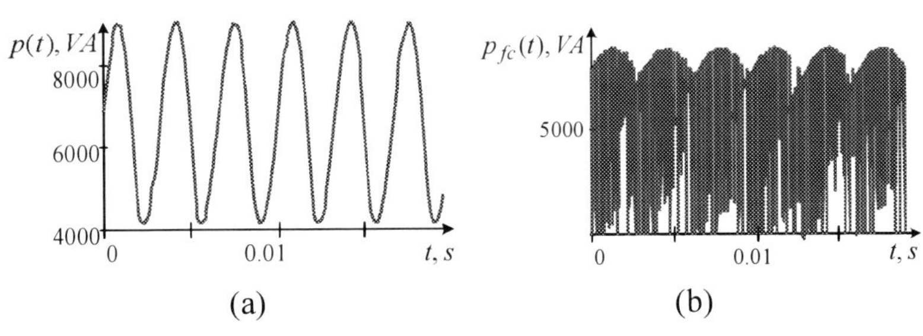

Figure 2.22. The signals of the active power consumed from the network (a) and from FC (b) by a symmetric IM.

The results of the research show that variable components are present in the signals of the consumed power even when operating with a symmetric IM. This is associated with non-sinusoidal currents consumed from the network and PWM voltage of the stator windings, which is formed by an autonomous inverter. Therefore, in the research of asymmetric modes of the operation of ED system with vector control, in the signals of power consumption, it is necessary to identify variable components caused by these factors, and variable components caused by the asymmetry of the stator windings. Thus, the current signals of the power supply network and the spectral composition of the active power consumed from the power supply network (shown without a constant component) during the operation of symmetric and asymmetric motors are presented in Figures 2.23-2.24.

Thus, in the operation of a variable-frequency ED with vector control of the IM with asymmetric stator windings, there is a distortion of the signal of the network current, and in the power spectrum the 2nd harmonic with a frequency of 100 Hz appears. In turn, the presence of the 6th harmonic with a frequency of 300 Hz is due to non-sinusoidal current signals.

In turn, the spectral composition of the active power consumed by the motor from the frequency converter without a constant component is presented in Figure 2.25.

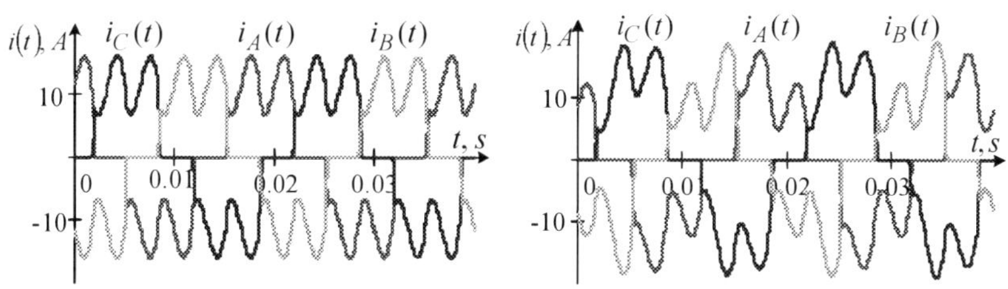

(a) (b)

Figure 2.23. Power supply current signals during operation of symmetric (a) and asymmetric (b) motor.

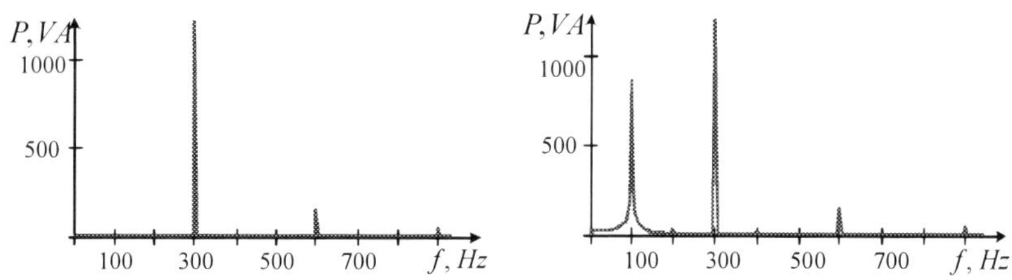

(a) (b)

Figure 2.24. Spectral composition of instantaneous power consumed from the network by symmetric (a) and asymmetric (b) motor.

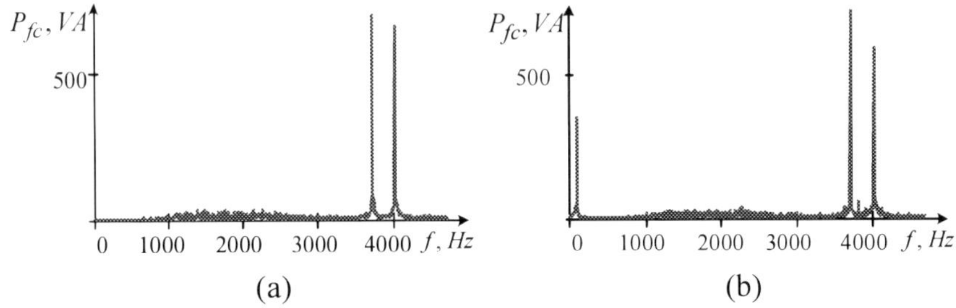

Figure 2.25. The spectral composition of the active power consumed from FC by a symmetric (a) and an asymmetric (b) motor.

The results of the research show that in the operation of an ED with an asymmetric motor, a harmonic with a frequency of 100 Hz also appears in the spectral composition of the power consumed from FC, in addition to harmonics caused by PWM voltage.

Thus, the 2nd harmonic in the signals of power consumption from both the supply network and FC is caused by the asymmetry of the motor windings. Therefore, harmonics above the 5th were excluded from the analysis of the variable components in the instantaneous power signals. The signals of the active power consumed by asymmetric IM are given in Figure 2.26.

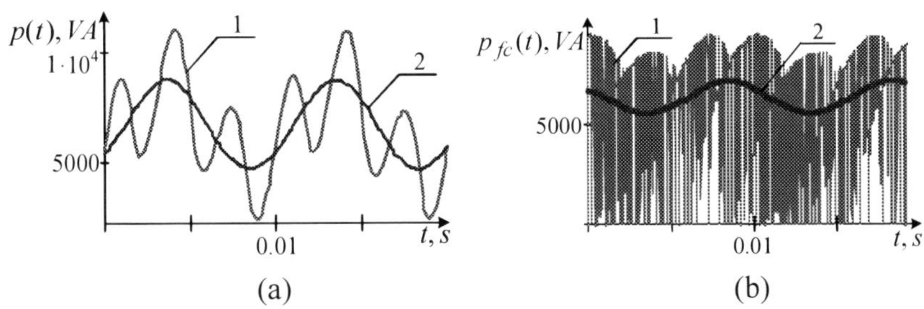

Figure 2.26. The signals of the instantaneous active powers consumed by an asymmetric motor from the network (a) and from FC (b): 1 – before filtration; 2 – after filtration.

The calculation of the variable component of the motor electromagnetic torque, by analogy with the variable component of power consumption, can be performed on the basis of expression:

$$\widetilde{T}_e(t) = T_e(t) - \frac{1}{T}\int_0^T T_e(t)dt, \tag{2.37}$$

where $T_e(t)$ – the electromagnetic torque of the motor.

The spectral composition of the electromagnetic torque of the motor (given without a constant component) during the operation of a symmetric and an asymmetric motor is presented in Figure 2.27.

Thus, a harmonic with a frequency of 100 Hz also appears in the spectrum of the electromagnetic torque of an asymmetric motor, in addition to harmonics at the frequency of PWM voltage. Therefore, when analyzing the instantaneous signals of the torque, harmonics above the 5th were also excluded. The signals of the electromagnetic torque of a symmetric and an asymmetric IM are given in Figure 2.28.

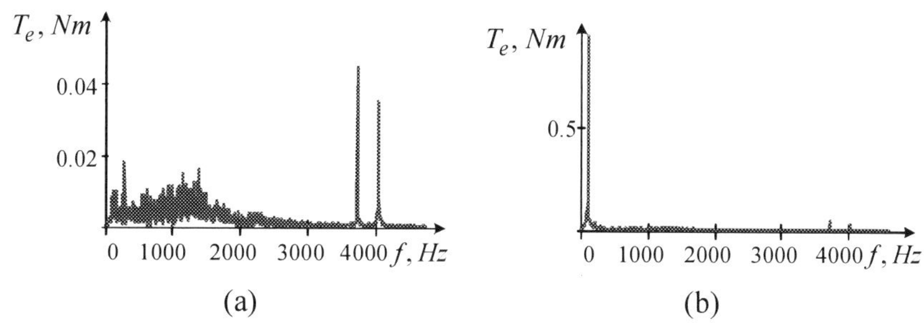

Figure 2.27. The spectral composition of the electromagnetic torque of a symmetric (a) and an asymmetric (b) motor.

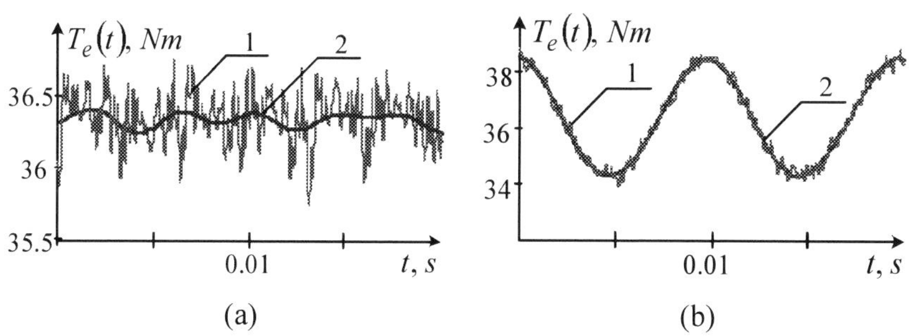

Figure 2.28. Electromagnetic torque of a symmetric (a) and an asymmetric (b) IM: 1 – before filtration; 2 – after filtration.

Conclusion

1. It has been determined that when using the known approaches to the research and construction of the vector control systems based on the use of orthogonal models of motors, it is impossible to take into account the asymmetry of motor windings, consequently, the researched object should be represented in three-phase coordinate system.
2. It has been experimentally proved that the mathematical model of an induction motor in a three-phase coordinate system taking into account the asymmetry of the stator windings makes it possible to adequately reproduce the electromagnetic and electromechanical processes occurring in asymmetric motors. Even with significant asymmetry, about 10–15% of the calculation error is up to 5% for idle mode and up to 1.5% when working with rated load.
3. It has been determined that when the induction motor stator windings are powered by a frequency converter, the increase in losses in the copper of the stator and rotor from higher harmonics of pulse-width modulated voltage is insignificant compared with the increase in losses caused by the asymmetry of stator windings. Therefore, the losses from higher harmonics can be neglected in the calculation.
4. The results of the research of the variable-frequency electric drive with vector control has shown that the asymmetry of the stator windings of the induction motor results in the redistribution of current by phase, which leads to significant overheating of individual windings even with a slight increase in total losses. In turn, asymmetric current loading of the motor phases causes local overheating in individual semiconductor switches of both the uncontrolled rectifier and the autonomous voltage inverter. Due to the fact that the energy loss of power keys is released in a semiconductor crystal and dissipated as heat, the long-term operation of the frequency converter in this mode can lead to significant overheating of individual switches, and subsequently to their complete failure.

Chapter 3

The Correction of the Operation Modes of Induction Motors with Stator Asymmetrical Windings during Scalar Control

3.1. The Development of a System to Compensate for the Impact of IM Asymmetry by Means of VFED

3.1.1. The Development of a Method for Compensating for the Influence of a Three-Phase IM Asymmetry by Means of VFED

The analysis of the instantaneous power theories revealed the possibility of using the provisions of the cross-vector theory of instantaneous power to compensate for the influence of IM stator windings asymmetry by means of VFED [167, 168]. There is no need to use an additional source of compensatory effects, as most industrial FCs are presented by converters with a DC link, which includes a self-excited voltage inverter (SVI). This unit can be not only a power source for the IM, but also a compensator. That is, based on the existing methods of calculating the compensatory effects it is possible to generate a voltage of a specific shape at SVI outputs, taking into account that an AC electric motor is the object of compensation [165, 169].

To compensate for the influence of IM asymmetry, the method of calculating the reference signals using the cross-vector theory of instantaneous power was chosen as the basic one [171, 172]. It involves the use of real, without intermediate conversion, current and voltage signals, which greatly simplifies the mathematical calculations in the control system. The disadvantage of this power theory consists in the fact that it does not control the currents in the neutral wire regardless of the phase currents and does not provide for full compensation for the current in the neutral wire if the voltages contain zero sequence. However, this is acceptable, because most often the stator windings, especially in the medium and high power IMs, are connected according to the scheme "star without zero" [173].

According to cross-vector theory, the reference currents on the compensator are defined as:

$$\vec{\mathbf{i}}^{*}_{cs} = \frac{\tilde{p} \cdot \vec{\mathbf{u}}_s}{u \cdot u} + \frac{\vec{\mathbf{q}} \times \vec{\mathbf{u}}_s}{u \cdot u}, \tag{3.1}$$

where $\tilde{p}$ – a variable component of three-phase active power consumed by IM; $\vec{\mathbf{u}}_s = [u_A \; u_B \; u_C]^T$ – the vector of the stator voltages; $u = \sqrt{u_A^2 + u_B^2 + u_C^2}$ – the modulus of the voltage vector of IM stator, $\vec{\mathbf{q}} = [q_A \; q_B \; q_C]^T$ – the vector of the instantaneous reactive power.

If IM is the object of compensation, the problem of reactive power compensation is not raised, at least for the first harmonics of voltages and currents. The primary task is to eliminate the variable component of three-phase active power and electromagnetic torque, but the research has shown that taking into account the reactive component will stabilize IM torque in compensation. To solve this problem by means of VFED it is necessary to calculate specific forms of three-phase voltage curves. It should be noted that the sequence of calculation of the correcting signals is the same for the case of compensation for the variable component of the three-phase active power, and for the compensation for the variable component of the torque. The only difference consists in the input signals of the control system: for the first case it will be the instantaneous values of currents and voltages of the phases of IM stator, and for the second case – the instantaneous values of flux linkages and currents of the stator phases.

3.1.1.1. The Compensation for the Variable Components of IM Instantaneous Power

Using (3.1), one can similarly write an expression for calculating the corrective voltage vector for the case when the variable components of IM instantaneous active and reactive power, which is caused by the asymmetry of its parameters, must be compensated for:

$$\vec{\mathbf{u}}^{*}_{ps} = \frac{\tilde{p}}{i} \frac{\vec{\mathbf{i}}_s}{i} + \frac{\vec{\mathbf{q}} \times \vec{\mathbf{i}}_s}{i \cdot i}, \tag{3.2}$$

where $\vec{\mathbf{i}}_s = [i_A \; i_B \; i_C]^T$ – stator current vector, $i = \sqrt{i_A^2 + i_B^2 + i_C^2}$ – stator current vector modulus.

In expression (3.2) components $\tilde{p}/i$ and $\vec{\mathbf{q}}/i$ determine the desired amplitude values of the harmonics of the corrective voltage, and component $\vec{\mathbf{i}}_s/i$ – the relative value of the stator phase current, which determines the frequency and phase of the compensating voltage of IM corresponding phase.

Adding corrective voltages calculated by expression (3.2) to the system of basic three-phase symmetric voltages of the rated amplitude and frequency allows obtaining such a voltage on the stator phase windings that will lead to the formation of instantaneous power without the variables caused by the asymmetry of currents in the stator phases [174].

The compensation for the variable component of the electromagnetic torque. By analogy with the three-phase power in the electromagnetic torque one can also distinguish "active" $\tilde{T}_{ep}$ and "reactive" $\vec{\mathbf{T}}_{eq} = \left[T_{eqA}\, T_{eqB}\, T_{eqC}\right]^T$ components. Thus, similar to expression (3.2), it is possible to calculate the voltage correction signals for the case when the variable component of the electromagnetic torque must be compensated for:

$$\vec{\mathbf{u}}^*_{Tes} = \frac{\tilde{T}_{ep}}{i}\frac{\mathbf{i}_s}{i} + \frac{\vec{\mathbf{T}}_{eq} \times \vec{\mathbf{i}}_s}{i \cdot i}. \tag{3.3}$$

The electromagnetic torque is calculated by expression:

$$T_e = \frac{p_n}{\sqrt{3}}\left[(\Psi_C - \Psi_B)i_A + (\Psi_A - \Psi_C)i_B + (\Psi_B - \Psi_A)i_C\right].$$

For implementing the offered method the generalized functional scheme of VFED with the system of compensation for the influence of asymmetry of three-phase loading is offered (Figure 3.1). The blocks denote the following: FC – frequency converter, R – three-phase rectifier, DCL – direct current link, SVI – self-excited voltage inverter, VS and CS – voltage and current sensors, SC – system of control, u_{cA}, u_{cB}, u_{cC} – correcting voltages.

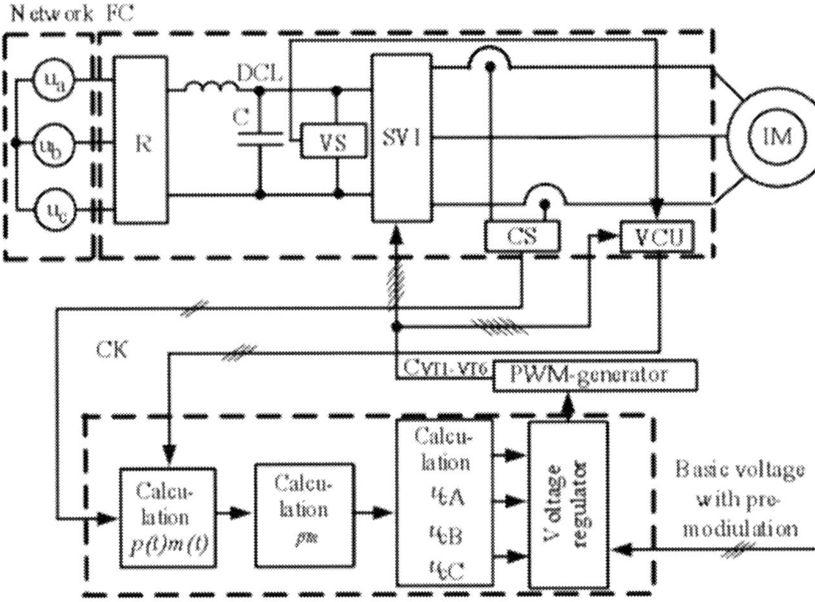

Figure 3.1. The generalized block diagram of VFED with a system of compensation for the influence of asymmetry of three-phase loading.

3.1.2. Mathematical Modeling of the Developed System of Compensation for the Influence of the Asymmetry of Three-Phase Loading by Means of VFED

The principle of operation of VFED SC (Figure 3.2) to compensate for the impact of load asymmetry consists in the following: three-phase active power (Figure 3.3) and reactive power (Figure 3.4) are calculated based on the instantaneous voltage and current signals measured by sensors and are decomposed into constant and variable components. After that, expressions 3.2-3.3 are used to calculate the reference voltages for each phase, which are supplied to SVI control input. It should be noted that in this scheme, the formation of voltage for each phase of the load occurs separately and independently of others [175].

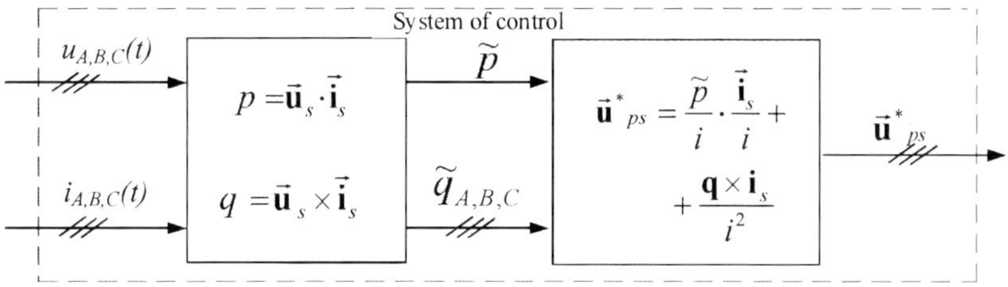

Figure 3.2. The block diagram of the system of control of the compensation for the variable component of three-phase power by means of VFED.

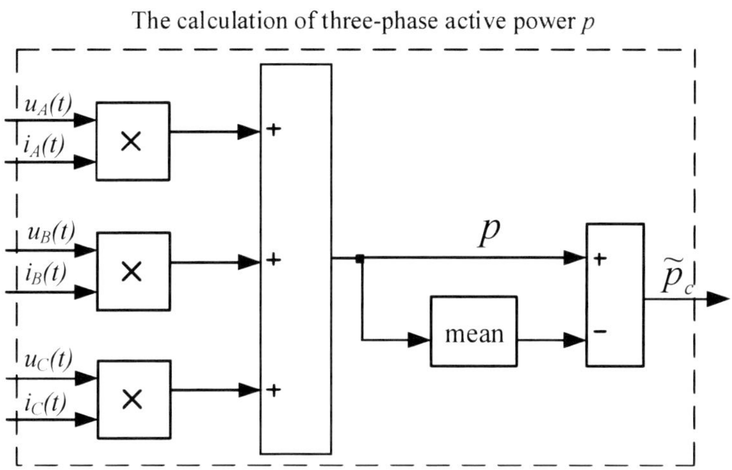

Figure 3.3. The block diagram of the active power calculation unit.

Mathematical modeling of modes of compensation for the influence of IM asymmetry by means of VFED according to the proposed scheme of SC structure is carried out with the use of structural modeling. The adequacy of the asymmetric IM model is experimentally confirmed in paper [176]. However, in modeling the researched system of ED with the proposed SC (Figure 3.2) there are a number of issues that need to be preliminarily clarified with other, simpler, models or models using other principles of modeling. Such issues in the use of structural modeling include the difficulty of taking into account the following features of the real power equipment:

- the distortion of the shape of the output voltage of the converter due to the PWM voltage;

- system performance at load connections according to the "star with zero" and "star without zero" circuits;
- the shift of "zero" at asymmetry of three-phase loading;
- limiting the maximum output voltage of the converter;
- the possibility of forming only linear control voltages with the help of real SVI, while the SC provides for the calculation and formation of the reference phase voltage on the SVI separately for each phase of the load [177, 178].

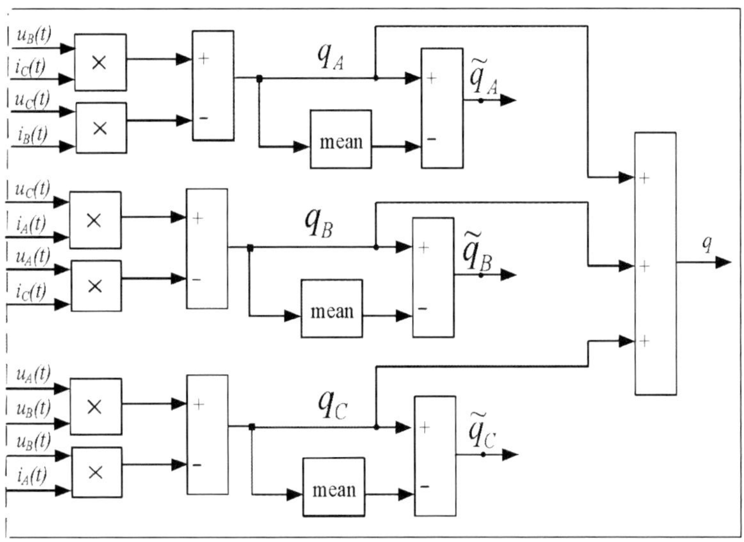

Figure 3.4. The block diagram of the reactive power calculation unit.

Taking into account the above, to verify the adequacy of the structural model of the proposed SC, the obtained results were compared with the results of modeling the researched system based on virtual units describing the operation of power electrical elements taking into account the above features. The advantage of the latter modeling method consists in the fact that complex electrical systems can be researched using the possibilities of both structural and virtual modeling [179]. In the researched three-phase system the power part is represented by means of virtual blocks, and the system of control – by means of blocks of structural modeling which reflect the algorithm of its operation.

3.1.2.1. Modeling the Operation of VFED with the System of Compensation for the Influence of the Asymmetry of Three-Phase Active-Inductive Loading

The analysis and comparison of the results of the structural and virtual modeling of the proposed system of compensation for the impact of load asymmetry by means of VFED at the first stage were performed using the example of modeling the operation of asymmetric three-phase active-inductive load with connection of windings according to the "star without zero" circuit supplied by alternating current through FC with a link of a direct current:

$$\begin{cases} u_A(t) = R_A \cdot i_A(t) + L_A \cdot \dfrac{di_A(t)}{dt}; \\ u_B(t) = R_B \cdot i_B(t) + L_B \cdot \dfrac{di_B(t)}{dt}; \\ u_C(t) = R_C \cdot i_C(t) + L_C \cdot \dfrac{di_C(t)}{dt}; \\ u_0(t) = \dfrac{1}{3}\left(i_A(t) + i_B(t) + i_C(t)\right)R_e, \end{cases} \tag{3.5}$$

where $u_A(t), u_B(t), u_C(t)$ and $i_A(t), i_B(t), i_C(t)$ – load phases voltage and currents; R_A, R_B, R_C – phases resistances; L_A, L_B, L_C – phases dissipation inductances; $u_0(t)$ – the voltage in the zero wire; $R_e = \mathrm{Re}(Z_e)$ – load equivalent resistance; $Z_e = 1/Y_e$ – load equivalent complex resistance; $Y_e = Y_A + Y_B + Y_C$ – load equivalent conductivity; $Y_A = 1/Z_A$; $Y_B = 1/Z_B$; $Y_C = 1/Z_C$ – phases conductivities; Z_A, Z_B, Z_C – phases complex resistances.

The first model of asymmetric RL-load operation was implemented on the basis of the principles of structural modeling (Figure 3.5).

In this circuit, the phase compensation voltages, which are calculated by the control system, are added, using an adder, to the basic three-phase voltage separately for each phase of the load.

Figure 3.6 shows the structure of the circuit model with virtual units that simulate the operation of a three-phase voltage source, bridge diode rectifier, autonomous voltage inverter, voltage and current sensors and three-phase active-inductive load winding.

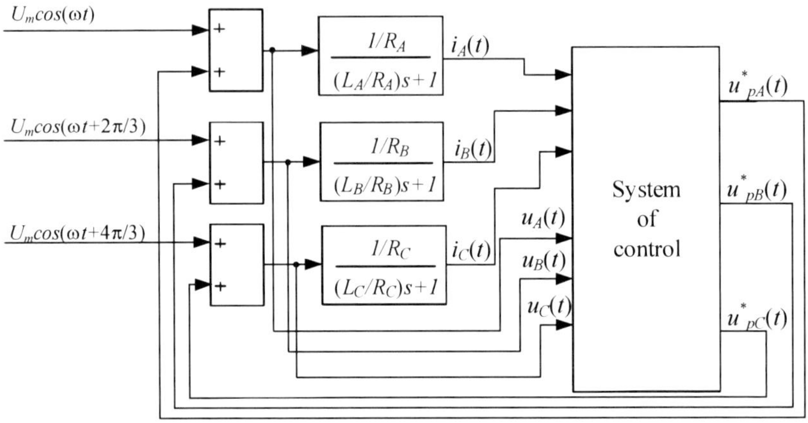

Figure 3.5. The block diagram of the SC of the model of compensation for the influence of three-phase load asymmetry, implemented using structural modeling (Circuit 1).

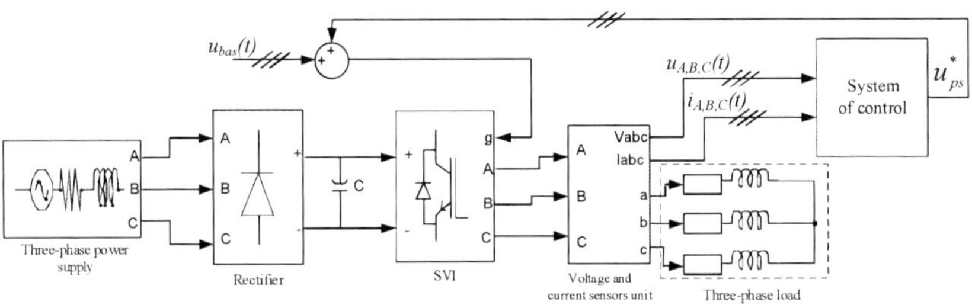

Figure 3.6. The block diagram for circuit modeling of SC compensation for the influence of the asymmetry of three-phase loading implemented on the basis of virtual blocks of elements of power electronics (circuit 2).

To research the operation of the compensation system, modeling was performed under the condition of artificially introduced phase asymmetry: the active resistance of the phase was calculated by expression $R'_A = R_A(1-\varepsilon_w)$, where ε_w – the percentage of the introduced asymmetry. The results shown in this paper are presented for the following cases: asymmetry in phase *A* is 1% ($\varepsilon_{wA} = 0.99$) (*mode a*); asymmetry in phase *A* is 2.5% ($\varepsilon_{wA} = 0.975$) (*mode b*); asymmetry in phase *A* is 5% ($\varepsilon_{wA} = 0.95$) (*mode c*); asymmetry in phase *A* is 10% ($\varepsilon_{wA} = 0.9$) (*mode d*). The load parameters are given in Table 3.1.

Table 3.1. Load parameters

Parameter	Value
$R_A = R_B = R_C$, Ohm	0.728
$L_A = L_B = L_C$, Hn	0.0036

Modeling in the given block diagrams was carried out under the same conditions and with the same load parameters (Figure 3.7-3.8).

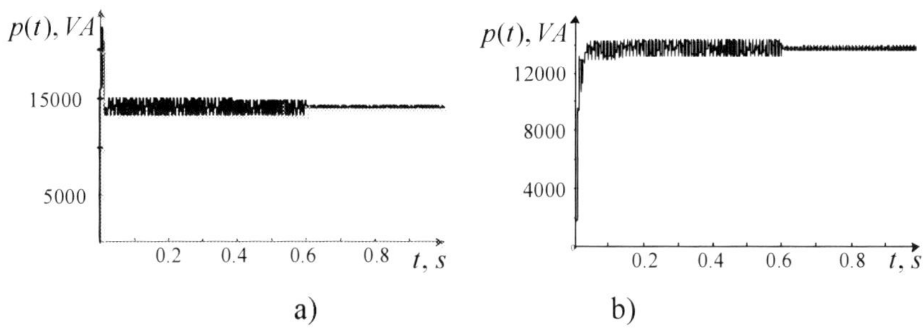

Figure 3.7. Three-phase active power on load with and without compensation for the asymmetry in phase A equal to 5% (*mode c*): a) according to circuit 1, b) according to circuit 2.

The modeling results revealed that after switching on the compensator, the variable component of the three-phase active load power is significantly reduced. Moreover, the results of the operation of the system of compensation for variable components of power are almost the same for both circuits of SC implementation, which confirms the adequacy of the structural model of the developed control system. Taking this into account, it is expedient to give the results of modeling SC operation for all the researched degrees of asymmetry by an example of only one circuit (Table 3.2).

As the modeling results demonstrate, the efficiency of the compensation system almost does not depend on the degree of asymmetry of the load parameters and remains quite high with its increase (the level of compensation for the variable component reaches 90%).

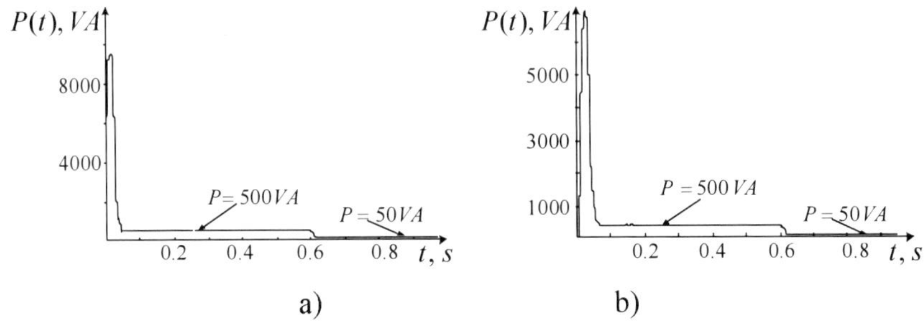

Figure 3.8. The effective value of the variable component of the three-phase active power with and without compensation for the asymmetry in phase A equal to 5% (*mode c*): a) according to circuit 1; b) according to circuit 2.

Table 3.2. The results of modeling SC operation

Parameter **Asymmetry degree**	**Active value of the variable power component** $\widetilde{p}(t)$, **VA**		**Relative value** $\widetilde{p}(t)/P_r(t)$, **%**	
	with	**without**	**with**	**without**
2.5 (*mode b*)	310	30	2.2	0.2
5 (*mode c*)	500	50	4	0.4
10 (*mode d)*	1200	105	8.5	0.8

When researching the operation of the compensation system it is necessary to control not only the variable component of the three-phase power load, but also the heating losses by phases, as asymmetry of parameters can lead to significant overheating of a single winding even with a small increase of total losses. Taking this into account, the calculation of load electrical losses with and without compensation for the asymmetry of parameters was performed.

Accordingly, the relative values of the change in losses in the power part of the variable-frequency ED can be calculated based on expression:

$$\Delta p = \frac{\Delta P(\varepsilon_W) - \Delta P_r}{\Delta P_r} \cdot 100\%, \tag{3.6}$$

where $\Delta P(\varepsilon_W)$ – the losses in ED elements for the current degree of asymmetry; ΔP_r – the losses in the corresponding elements of ED with a symmetric IM.

To obtain more complete and informative data, research was conducted for several variants of the initial inductance of the phases (cosφ = 0.95; cosφ = 0.85; cosφ = 0.75) and for two winding connections: “star with zero” and “star without zero wire. The results of the calculations are given in the form of diagrams, which show the total and phase electrical losses before and after turning on the compensator (Figure 3.9-3.11). The following symbols are accepted in the given figures: – relative values of the stator copper losses of phase A of the motor; – relative values of the stator copper losses of phase B of the motor; – relative values of the stator copper losses of phase C of the motor; – relative values of the total stator copper losses of the motor.

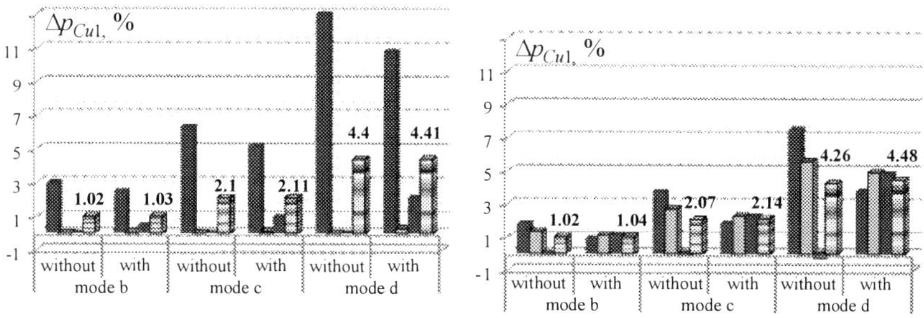

(a) (b)

Figure 3.9. Total and phase electrical losses of RL-load before and after switching on the compensator at cosφ = 0.95: a) for the circuits with a zero wire and b) for the circuit without zero wire.

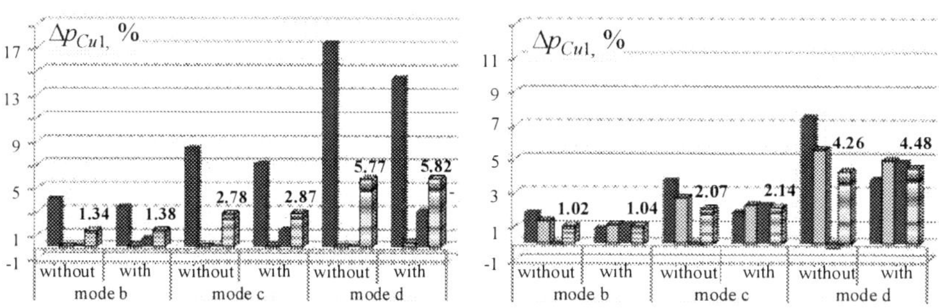

(a) (b)

Figure 3.10. Total and phase electrical losses of RL-load before and after switching on the compensator at cosφ = 0.85: a) for the circuits with a zero wire and b) for the circuit without zero wire.

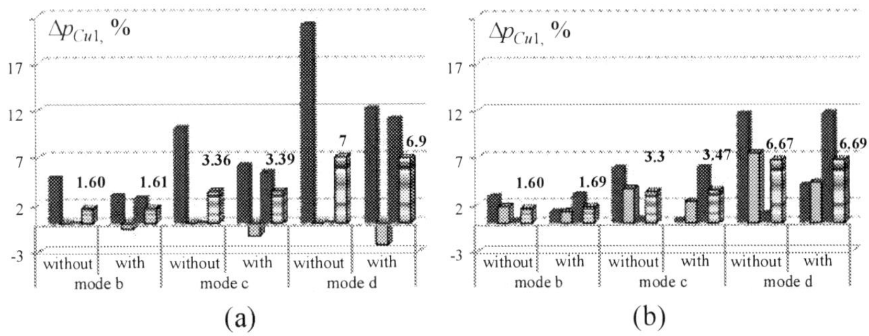

Figure 3.11. Total and phase electrical losses of RL-load before and after switching on the compensator at cosφ = 0.75: a) for the circuits with a zero wire and b) for the circuit without zero wire.

The above diagrams illustrate that as the degree of asymmetry increases, the total and phase electrical losses increase. However, if the total losses are within acceptable limits, the losses in the individual phases reach a fairly high level, which can lead to significant overheating of an individual phase winding of the load. Obviously, the use of a compensator allows not only reducing the variable component of three-phase power, but also redistributing losses in the load phases, while the total value of losses increases insignificantly. It is worth noting that in circuits without neutral wire the losses are distributed more evenly, and the losses in the load phase do not reach such large values as in the circuits with a neutral wire.

In addition, a comparative analysis of the results of structural (Figure 3.5) and virtual (Figure 3.6) modeling of the researched system showed that the proposed compensation system makes it possible to form compensation voltages not only with separate control of each phase of three-phase load, which is seldom possible in real systems of this type, but also in the joint control of all phases, as is the case in industrial VFED systems. These results allow us to move to the next stage of testing the effectiveness of the proposed control system – to model the operation of VFED system with the function of compensating for the impact of IM asymmetry.

3.1.2.2. Modeling of VFED Operation with the System of Compensation for the Influence of IM Asymmetry

The proposed method of compensating for the influence of motor asymmetry was verified by modeling the operation of the FC-AD system in a three-phase coordinate system at a rated load for different degrees of the stator phase asymmetry and for the cases of adjustment of the control system for the compensation for the variable component of three-phase active power and the variable component of the electromagnetic moment. The change in the resistance, inductive reactance of the winding and the change in the inductance of the magnetization circuit were taken into account during modeling the asymmetric modes [176, 180].

The block diagram of VFED with the function of compensating for the impact of IM asymmetry contains the same structural elements as the generalized block diagram of VFED to compensate for the impact of asymmetry for the three-phase load (Figure 3.12). The difference consists in the fact that the circuit below uses only two-phase current and voltage sensors, as the corresponding signals of the third phase can be calculated from the two measured signals:

$$i_B(t) = -(i_A(t) + i_C(t)).$$

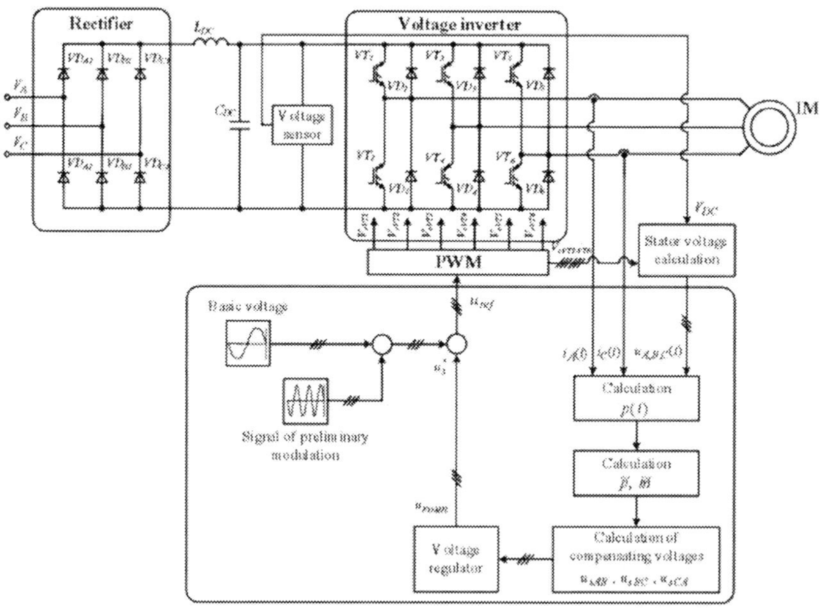

Figure 3.12. Functional diagram of VFED with the function of compensation for the influence of asymmetry of the induction motor.

According to the presented circuit, in this SC, the signals supplied to the pulse-width modulation (PWM) module are the sum of the reference basic three-phase voltage after its preliminary modulation and the system of compensation voltages calculated by 3.2 or 3.3 (Figure 3.13).

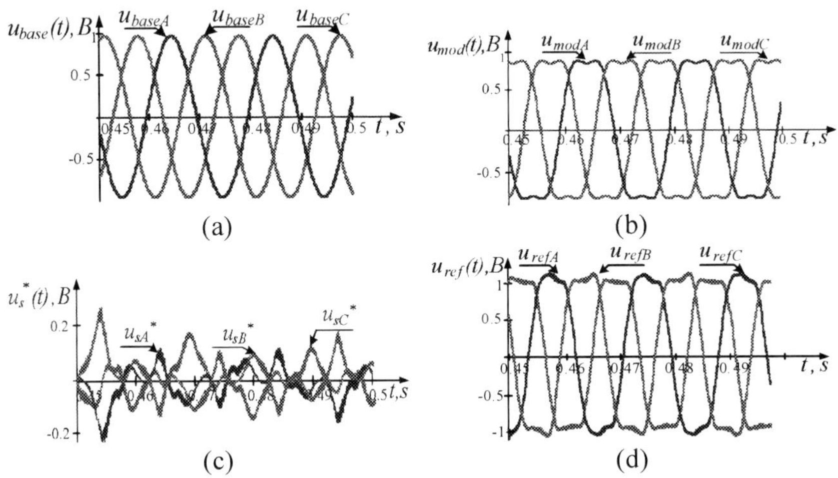

Figure 3.13. The control signals in SC: a) reference basic three-phase voltage; b) reference basic three-phase voltage with preliminary nodulation of the third harmonic; c) compensation voltages; d) PWM generator input signals, which are the sum of the reference basic voltage after preliminary modulation and compensation voltages.

The research of the modes of the operation of VFED with a scalar control at different degrees of asymmetry of IM stator windings was conducted for the motor. The obtained results (Figure 3.14-3.15) showed that after turning on the compensator, the RMS value of the variable component of the three-phase active power and, as a consequence, the variable component of the electromagnetic moment significantly reduces. For example, the results obtained for the case of 5% asymmetry of the stator windings, with no-load start and at adding the rated resistance moment, show that the RMS value of the variable component $\widetilde{T}_e$

the moment, and after turning on the compensator, it decreased to 1% (Figure 3.14).

The results illustrating the three-phase active power of IM before and after compensation also confirm the effectiveness of the proposed method of compensation (Figures 3.16-3.17).

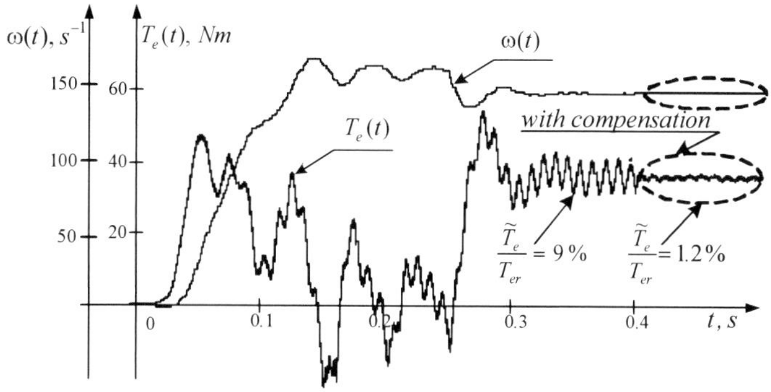

Figure 3.14. The electromagnetic torque and angular rotation frequency of IM at 5% phase asymmetry (mode c) before and after switching on the compensator when compensating for inactive components of the electromagnetic torque.

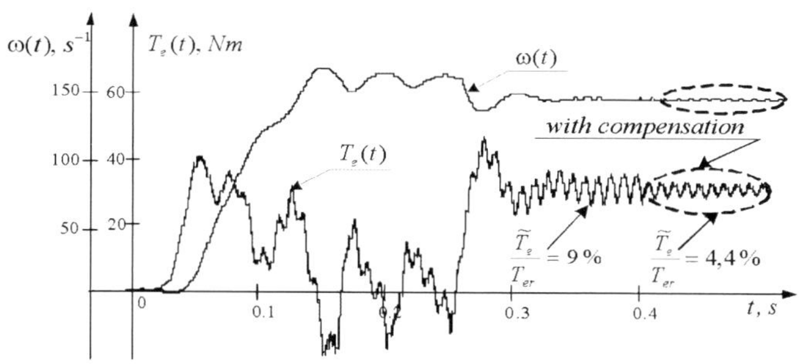

Figure 3.15. The electromagnetic torque and angular rotation frequency of IM at 5% phase asymmetry (*mode c*) before and after switching on the compensator when compensating for inactive components of the three-phase active power.

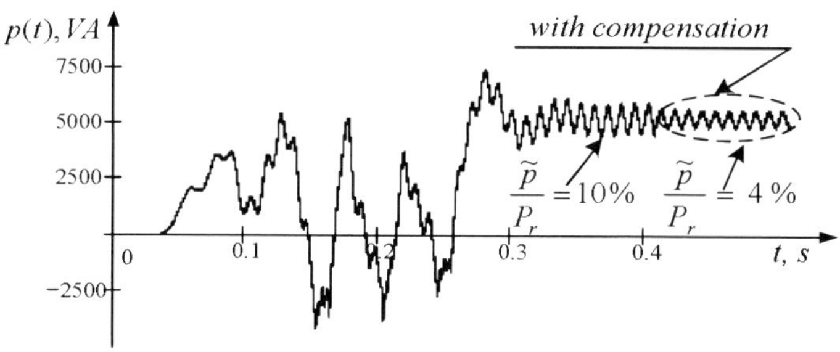

Figure 3.16. Three-phase active power of IM at 5% asymmetry of phases (*mode c*) before and after switching on the compensator at the compensation for the inactive components of the electromagnetic torque.

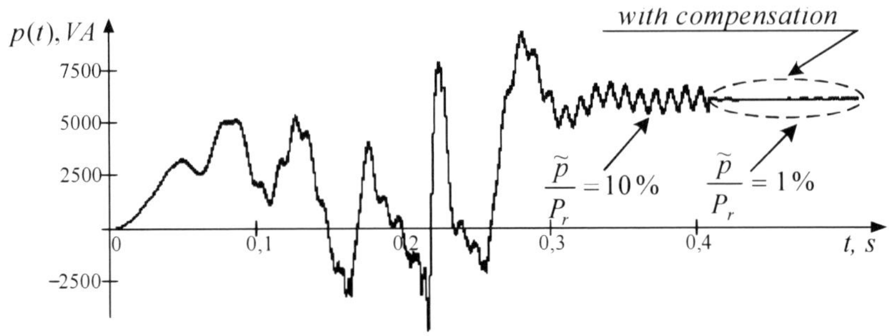

Figure 3.17. Three-phase active power of IM at 5% asymmetry of phases (*mode c*) before and after switching on the compensator at the compensation for the inactive components of the three-phase active power.

The analysis of the obtained results showed that when adjusting SC to compensate for the variable component of power, the variable component of the electromagnetic torque also decreases. However, the level of reduction of this component is smaller compared with the mode of setting the SC to compensate for the variable component of the electromagnetic torque (Figures 3.18-3.19). Taking this into account, if the priority is to eliminate the variable component in the electromagnetic torque of IM, the SC must be adjusted accordingly. In the given figures the following designations of the researched parameters are accepted:  – before compensation;  – after compensation.

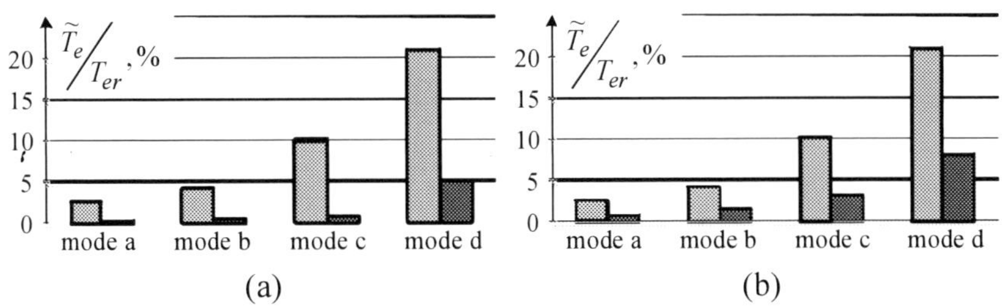

Figure 3.18. The relative value of the variable component of the electromagnetic torque of IM at different degrees of asymmetry before and after turning on the compensator when compensating for inactive components of the electromagnetic torque (a) and inactive components of three-phase active power (b).

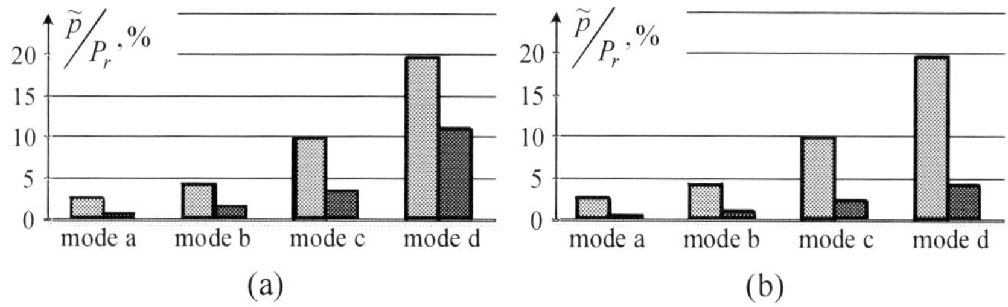

Figure 3.19. The relative value of the variable component of the three-phase active power of IM at different degrees of asymmetry before and after turning on the compensator to compensate for inactive components of electromagnetic torque (a) and inactive components of three-phase active power (b).

As the next stage of verifying the operation of the developed compensation system, modeling of the operation of IM with different degrees of multiphase asymmetry of SC stator was carried out (Table 3.3, Figures 3.20-3.21).

Table 3.3. The assessment of IM performance indicators before and after turning on the compensator for various cases of multiphase asymmetry

Phases asymmetry degree, %		No. 1*	No. 2**	No. 3*	No.4**	No. 5*	No. 6**
Indicator		A = 12.5 B = 6.25 C = 0	A = 10 B =3.3 C = 0	A = 0, B = 4.2 C=12.5	A = 16 B = 0 C =5.5	A = 16 B=12.5 C = 10	A=12.5 B = 6.2 C = 5.5
$\tilde{p}/P_r$, %	before	20	17	22	32	13	14
	after	2	4	6	22	2	6
$\tilde{T}_e/T_{er}$, %	before	21	17	23	33	12	15
	after	5,2	1	11	16	5	2
$\tilde{p}$, *VA*	before	1150	1000	1280	1800	735	830
	after	100	245	340	1340	105	380
$\tilde{T}_e$, *Nm*	before	7.2	6.1	8.1	11.5	4.5	5.2
	after	1.85	0.45	4	5.5	1.8	0.88

* – setting the SC to compensate for the variable components of IM instantaneous power;
** – setting the SC to compensate for the variable components of the electromagnetic torque.

The results of mathematical modeling proved the effectiveness of the proposed compensation method and being applied to the problem of compensation for the impact of IM asymmetry, they showed that turning on the compensator reduces the undesirable variable component of torque almost by an order, and, consequently, vibration reduces and ED operation conditions improve in general [181]. The developed block diagrams enable modeling ED power equipment work taking into account the features of their operation caused by the nature of processes of energy transformation and transfer.

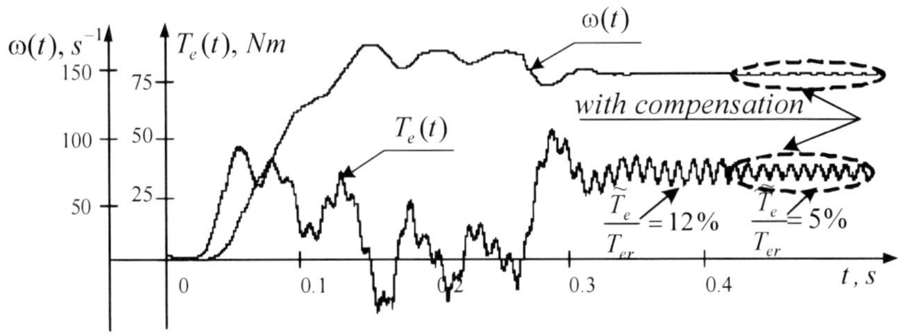

Figure 3.20. Electromagnetic torque and angular rotation frequency of IM with asymmetry in phase A equal to 10% and phase B equal to 3.3% ($\varepsilon_{wA} = 0.9$, $\varepsilon_{wB} = 0.967$) before and after switching on the compensator when compensating for inactive components of three-phase active power.

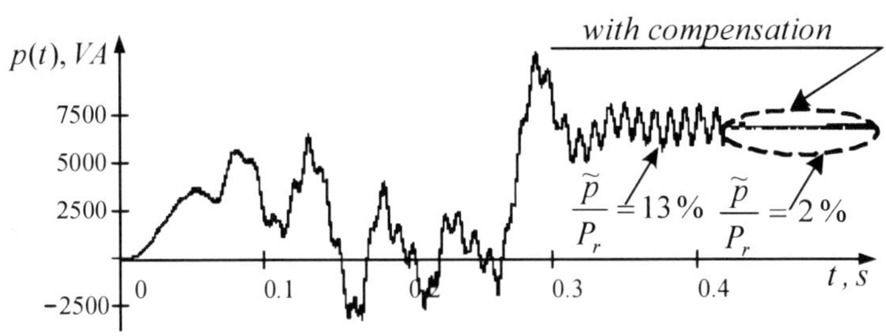

Figure 3.21. Three-phase active power of IM with asymmetry in phase A equal to 10% and asymmetry in phase B equal to 3.3% (εwA = 0.9, εwB = 0.967) before and after switching on the compensator when compensating for inactive components of three-phase active power.

3.2. The Assessment of the Stator Phase Asymmetry Influence on IM Service Life

Both in the research of the compensation system with asymmetric RL-load, and in the operation of VFED with asymmetric IM, it is necessary to control not only the variable component of three-phase load power but also heating losses by phases, as the asymmetry of parameters leads to the asymmetry of current loading of phase windings (Figure 3.22). This, in turn, is the cause of significant overheating of a separate winding, which, in the case of an induction motor, will reduce the service life of the entire IM.

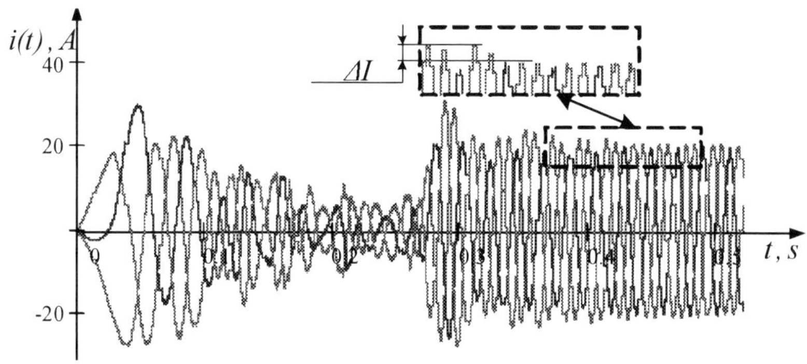

Figure 3.22. IM stator currents at asymmetry εwA = 5% before and after switching on the compensator when compensating for inactive components of the electromagnetic torque.

It is possible to assess quantitatively the phase asymmetry influence on IM service life using the rule of "eight degrees" [182-184]. To do this, it is necessary to determine the excess insulation temperature of the windings before and after turning on the compensator. It was assumed in the calculations that the insulation temperature of the windings varies in proportion to the heating losses of the motor.

The results of the research of electrical losses in the phases of the motor stator for the considered methods of compensation are given in Figures 3.23-2.24. The following notations of the researched parameters are adopted in the figure: ■ – the relative values of copper losses in the stator phase A of the motor; □ – the relative values of copper losses in the stator phase B of the motor; ◩ – the relative values of copper losses in the stator phase C of the motor.

The above diagrams show that as the degree of asymmetry increases, the total and phase electrical losses grow. However, while the total losses are within acceptable limits, the losses in some phases reach too high a level, which can lead to significant damage to the insulation of the winding due to overheating.

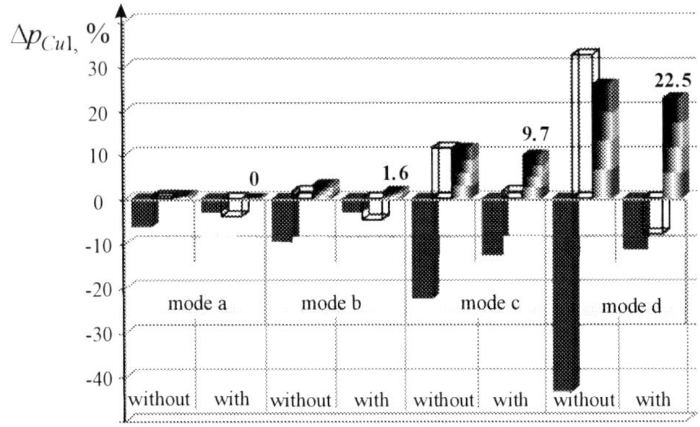

Figure 3.23. The relative value of the change in electrical losses in IM stator phases at different degrees of asymmetry before and after turning on the compensator to compensate for the inactive components of the electromagnetic torque.

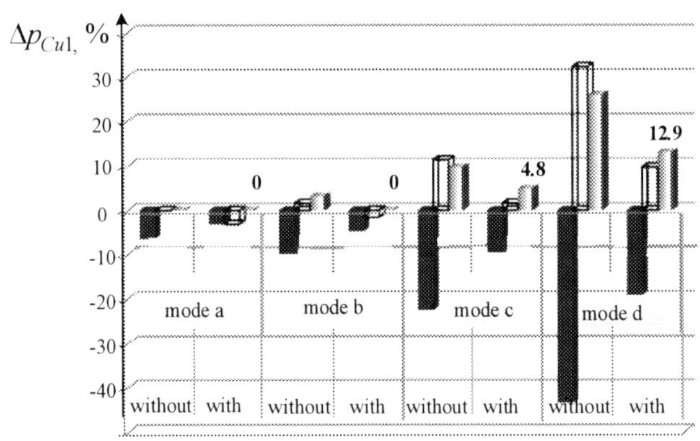

Figure 3.24. The relative value of the change in electrical losses in IM stator phases at different degrees of asymmetry before and after turning on the compensator to compensate for the inactive components of the three-phase power.

According to the results, before compensation phase B is the most overheated (at 10% asymmetry the electrical losses in the winding exceed the rated ones by more than 30%), so the IM residual life is calculated for the most overheated winding:

$$\tau(\varepsilon) = T_0 \cdot e^{-b \cdot \Theta(\varepsilon_w)}, \tag{3.7}$$

where $\Theta(\varepsilon_w) = \Theta_r \dfrac{\Delta P_{Cu1}(\varepsilon_w)}{\Delta P_{Cu1r}}$ – the temperature of the windings insulation at the given degree of asymmetry (when determining the excess temperature of the windings, one can use the ratio between the rated losses and the rated insulation temperature of the windings and assume that this factor is constant and does not depend on the mode of motor operation); Θ_r – the rated temperature of windings insulation, determined by IM insulation class; T_0 – the conditional service life of insulation at temperature $\Theta = 0$; b – the coefficient depending on the motor insulation class.

According to Figures 3.25-3.26 the change in IM service life after the compensation is calculated for each degree of asymmetry for the cases of adjustment of SC to compensate for the variable components of three-phase power (Table 3.4) and electromagnetic torque (Table 3.5).

Table 3.4. IM service life at the compensation for the variable components of three-phase power

No.	Asymmetry degree, %	Exceeding the insulation temperature of the most overheated winding, °C		Insulation service life τ, years	
		before	after	before τ_1	after τ_2
1	1 (mode a)	105.8 (phase B)	105 (phase C)	6.6	7
2	2 (mode b)	108 (phase C)	105 (phase C)	5.2	7
3	5 (mode c)	117 (phase C)	110 (phase C)	2.5	4.5
4	10 (mode d)	139 (phase B)	119 (phase C)	0.37	2.2

Table 3.5. IM service life at the compensation for electromagnetic torque variables

No.	Asymmetry degree, %	Exceeding the insulation temperature of the most overheated winding, °C		Insulation service life τ, years	
		before	after	before τ_1	after τ_2
1	1 (mode a)	105.8 (phase B)	105 (phase C)	6.6	7
2	2 (mode b)	108 (phase C)	107 (phase C)	5.2	6.05
3	5 (mode c)	116 (phase C)	115 (phase C)	2.5	2.9
4	10 (mode d)	139 (phase B)	128 (phase C)	0.37	0.9

Obviously, the use of a compensator makes it possible to redistribute the losses in the load phases [119], which, in turn, increases the service life of the insulation of the windings and the motor as a whole. In addition, such redistribution of losses and balancing of current amplitudes in the stator phases of the motor, in turn, reduces the current load of the corresponding arm of the inverter and, consequently, reduces the thermal overload of its keys.

From the point of view of the reduction of the thermal overload of separate phases the operation of VFED with control system adjustment to the compensation for the variable components of three-phase active power demonstrates the best results, but at the same time the level of compensation of the variable component of the electromagnetic torque decreases. Taking this into account, the compensation strategy should be chosen in view of the requirements for the operation of a particular ED of the working mechanism.

3.3. The Research of the Losses in the Power Semiconductor Keys of the Autonomous Voltage Inverter with the Compensation for IM Asymmetry Influence

As shown in Section 3.2, the asymmetry of the stator windings leads to current overload of IM individual phases. However, if such a motor is part of VFED, the specified mode of operation will also result in the redistribution of currents in FC power switches feeding the windings of the motor phases. The technical documentation for most industrial FCs indicates their ability to withstand maximum current overload (130-150%) for a short time (60 s according to the operating instructions of the frequency converter SINAMICS G120D), and 10% asymmetry of IM stator leads to current overload of individual phases by 30%. Taking this into account, the research included the IM asymmetry influence on the thermal state and operation modes of power semiconductor switches, which are an integral part of FC.

The calculation of electrical losses in SVI power switches before and after compensation was performed for different cases of single- and multiphase asymmetry and for different cases of setting up the compensation system: the mode of setting the SC to compensate for the variable component of the three-phase power consumption is marked "compensation $\widetilde{p}$," and the compensation mode of the variable component of the electromagnetic torque – "compensation $\widetilde{T}_e$" (Figure 3.25-3.29) [185]. The parameters of the power keys are given in Chapter 2.

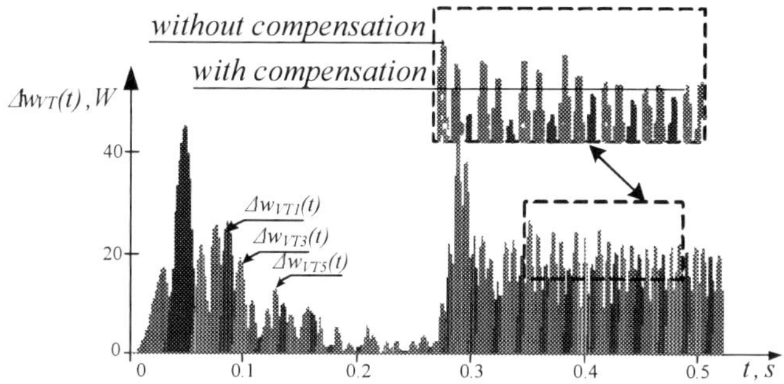

Figure 3.25. Power loss in power transistors VT1, VT3, VT5 with asymmetry in phase A equal to 5% and phase C – equal to 3% ($\varepsilon_{wA} = 0.95$, $\varepsilon_{wC} = 0.97$) before and after compensation.

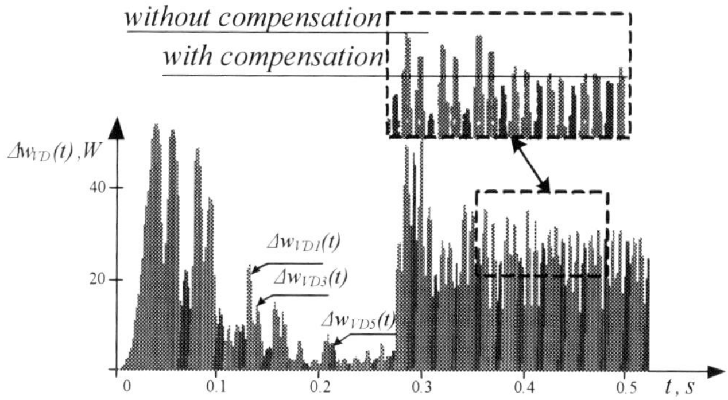

Figure 3.26. Power loss in power diodes VD1, VD3, VD5 with asymmetry in phase A equal to 5% and phase C – equal to 3% ($\varepsilon_{wA} = 0.95$, $\varepsilon_{wC} = 0.97$) before and after compensation.

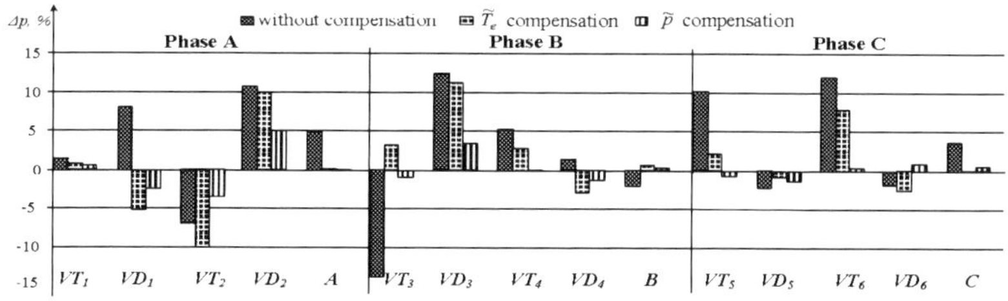

Figure 3.27. The deviation of power losses from losses in symmetrical mode in power transistors and SVI diodes with asymmetry in phase A equal to 5% ($\varepsilon_{wA} = 0.95$) before and after compensation.

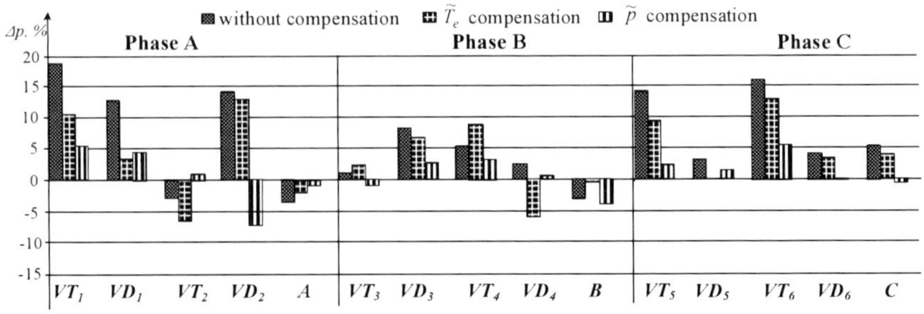

Figure 3.28. The deviation of power losses from losses in symmetrical mode in power transistors and SVI diodes with asymmetry in phase A equal to 5% and phase C – equal to 3% (ε_{wA} = 0.95, ε_{wC} = 0.97) before and after compensation.

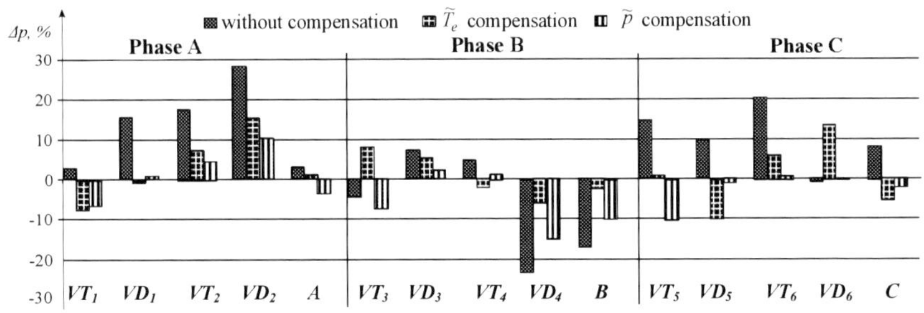

Figure 3.29. The deviation of power losses from losses in symmetrical mode in power transistors and SVI diodes with asymmetry in phase A equal to 10% ans phase C – equal to 7% (ε_{wA} = 0.9, ε_{wC} = 0.93) before and after compensation.

According to the results of the research, the asymmetry of IM stator windings negatively affects the mode of operation of SVI power semiconductor switches, as asymmetric current loading of IM phase leads to asymmetric distribution of currents flowing through transistors and the inverter reverse diodes. Long-time increased direct currents result in a significant excess of losses in individual SVI switches, and the reverse diodes of the branch of the bridge circuit, which is connected to IM damaged phase, are the most overloaded. Thus, when the asymmetry of one of the phases is 5%, another – 3%, the power loss in some reverse diodes exceeds the rated ones by almost 30%, and the power loss on the transistors – by almost 20%. The analysis of the calculated losses with and without compensation for the asymmetry of IM stator windings shows that the value of losses in the most overloaded inverter switches significantly reduces due to the compensation, and insignificantly increases in underloaded switches, approaching the value of losses in a symmetric IM. Thus, there is redistribution of currents, and, as a consequence, losses in SVI valves. It should be noted that, in terms of reducing the thermal overload of individual switches, the best results are shown by ED operation with the control system adjusted to compensate for variable components of power consumption, but in this case the level of compensation for the variable components of electromagnetic torque reduces.

Thus, it is proved that the application of the system of compensation for the influence of IM asymmetry makes it possible to improve the operating conditions of the semiconductor converter due to the redistribution of losses on the power switches of the autonomous voltage inverter after compensation.

3.4. The Calculation of the Voltage Regulator in the System of Compensation for IM Asymmetry Influence by Means of VFED

The generalized block diagram of VFED with the system of compensation for IM influence contains a block of control voltage regulation, which is a proportional regulator and determines the amplitude of compensatory voltage as the form and harmonic structure of this voltage is calculated in SC (Figure 3.12). Therefore, in order to properly adjust the SC to compensate for the variable components of the three-phase active power or IM electromagnetic torque, it is necessary to calculate the parameters of the P-regulator, i.e., the gain coefficient in SC direct channel. However, the analytical calculation of this coefficient is quite a difficult task, as it depends not only on the desired level of compensation of the variable component of IM original coordinate, but also on the parameters of the motor: the degree of IM asymmetry and the level of its load. The solution to this problem is possible using the methods of experimental planning theory, i.e., it is necessary to obtain VFED system regression model with the function of compensating for the impact of IM asymmetry, in which the gain coefficient in SC is one of the influencing factors.

Thus, to determine the type and calculation of the coefficients of the regression equation, which would mathematically describe the physical dependence of the source (either dependent on the input or independent) coordinates with sufficient reliability, it is necessary to conduct a series of experiments on the developed mathematical model. As the method of compensation for a variable component of IM three-phase power caused by

IM asymmetry is researched in this paper, the object of the research is presented by VFED with an asymmetric IM, the original (dependent) coordinates of which are the variable component of IM three-phase active power and the variable component of the electromagnetic torque (Figure 3.30).

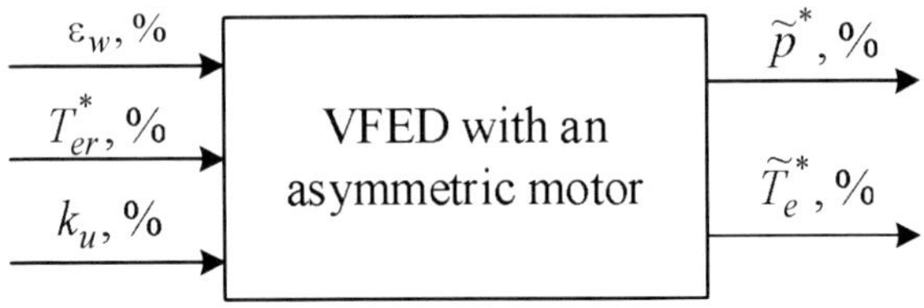

Figure 3.30. The research object, where: $\varepsilon_w, \%$, $T_{er}^*, \%$, $k_u, \%$ – input parameters, subject to control; $\tilde{p}^*, \%$, $\tilde{T}_e^*, \%$ – optimization parameters.

According to IM mathematical model in a three-phase coordinate system, the level of the variable component of three-phase power and, accordingly, IM electromagnetic torque in a closed control system are to be affected by the following: the degree of asymmetry or non-sinusoidality of the input voltage, the degree of asymmetry of the motor phase windings and the moment of resistance on IM shaft. In addition, a significant factor in the impact is the gain coefficient in SC direct channel, which is implemented on a three-phase active motor power.

Thus, it is possible to list the factors that should be taken into account in the model of compensation for the impact of IM asymmetry by means of VFED: X_1 – the degree of asymmetry of motor phase windings $\varepsilon_w, \%$; X_2 – the moment of resistance on IM shaft $T_{er}^*, \%$; X_3 – the gain coefficient in SC direct channel $k_u, \%$. It should be noted that this model is not used to research the effectiveness of compensation for the variable component of IM three-phase active power caused by asymmetry or non-sinusoidal supply voltage, as at the present stage there is a significant number of publications by other authors [186-188], whose research is devoted to this issue. Therefore, the verification of the influence of these factors requires a complete factorial experiment (CFE), which would take into account all possible combinations of the influence of these factors. Hypothesis H_0 – the amplitude of the variable component of IM three-phase active power in a closed SC is influenced by the degree of asymmetry of IM phases windings, the moment of resistance on IM shaft and the gain coefficient in the direct channel. Alternative hypothesis

H_1 – the amplitude of the variable component of IM three-phase active power in a closed SC does not depend on factors X_1, X_2, X_3.

When varying factors at two levels, it is necessary and sufficient to carry out $N = 2^k$ experiments, where 2 – the number of levels; k – the number of factors, i.e., in this paper the number of experiments is equal to $2^3 = 8$, and the number of repetitions – 10.

The creation of the CFE plan is reduced to the choice of the experimental point, which is located symmetrically relative to the zero level. Encoded values are entered to simplify writing:

"+" – upper level;
"-" – lower level.

These values are based on the following transformation:

$$u_i = \frac{x_i - x_{0i}}{h_i},$$

where u_i – the encoded value of the factor; x_i – the natural value of the factor; x_{0i} – the value of the factor at the center of the plan; h_i – the natural value of the variation interval; i – the number of the factor [189].

Thus, to conduct a complete factor experiment for each of the factors it is necessary to determine the "zero point" and symmetric experimental points relative to it, and it should be noted that too wide a range of factors will reduce the accuracy of the resulting regression model, and too narrow one will limit the practical application of the resulting model. Therefore, the degree of the asymmetry of the stator phases is taken as "zero point" $\varepsilon_w = 8\%$ for factor X_1, as at such value of asymmetry the efficiency of IM operation essentially worsens [31, 190]; for factor X_2 – the relative value of the moment of resistance on the shaft $T_{er}^{*} = 85\%$, as a large number of motors operate in the mode of insignificant underload [22]; for factor X_3 the value of the gain coefficient is taken according to preliminary model experiments $k_u = 20$ (Figure 3.14-3.15). Table 3.6 shows the designations, names and levels of variation of independent factors of experiments in the research of SC operation to compensate for the impact of IM asymmetry by means of VFED:

A preliminary plan of the complete factor experiment was created for a full-scale research of these factors influence on the level of the variable component in IM electromagnetic torque. Its structure is presented in Table 3.7.

Table 3.6. The list and levels of the factors change

No.	Factor	Variation levels	
		-1	1
1	The degree of asymmetry of IM stator windings, ε_w, %	4	12
2	The relative moment of resistance on IM shaft, T^*_{er}, %	70	100
3	Gain coefficient in the direct channel, k_u	15	25

Table 3.7. The plan of the complete factor experiment

№	X_1	X_2	X_3	X_1X_2	X_1X_3	X_2X_3	$X_1X_2X_3$
1.	-1	-1	-1	1	1	1	-1
2.	-1	-1	1	1	-1	-1	1
3.	-1	1	-1	-1	-1	1	1
4.	-1	1	1	-1	1	-1	-1
5.	1	-1	-1	-1	1	-1	1
6.	1	-1	1	-1	-1	1	-1
7.	1	1	-1	1	-1	-1	-1
8.	1	1	1	1	1	1	1

The specifics of the model experiment consist in the following:

1. In VFED closed system IM was modeled according to a system of equations in three-phase coordinates, which took into account the asymmetry of the electromagnetic parameters of the stator and the motor supply from the SVI.
2. The experiments were performed for the case of adjusting the SC to compensate for the variable component of IM electromagnetic torque (expression 3.3).
3. Each experiment was performed 10 times; the arithmetic mean was entered as a result of the experiment.
4. The relative value of the variable component of IM electromagnetic torque torque is chosen as the optimization parameter, since the electromagnetic moment is the most important and informative

parameter that characterizes the operation of the electric drive in general.

5. The plan of experiment 2^3 was used to create dependences $\tilde{T}_e^* = f(\varepsilon_w, T_{er}^*, k_u)$ and $y = f(x_1, x_2, x_3)$.
6. Processing of experimental results was performed using MS Excel and StatGraphics Centurion XV packages.

Regression coefficients are determined by formula:

$$b_i = \sum x_{iu} y_u / n \quad (3.8)$$

where x_{iu} – the value of the factor, y_u – the result of the experiment, N – the number of experiments.

In the course of statistical analysis of the experimental results, the assessment of the mathematical expectation and variance of the general totality was performed according to the following expressions [191]:

$$M(x) = \bar{y} = \sum y_i / n \quad (3.9)$$

$$D(x) = \sigma^2 = \frac{1}{n-1} \sum (y_i - \bar{y})^2 \quad (3.10)$$

According to the requirements of regression analysis, the use of the results of experimental research is possible only in the case of equality of variances of response functions at each point of the experiment (homogeneity of variances). The verification of the homogeneity of variances was performed on the basis of their statistical assessments (3.11) by the Cochran test (G – test) $H_o : G \le G_{cr}$ [189]:

$$G = \frac{\sigma_{\max}^2}{\sum_{i=1}^{n} \sigma^2} \quad (3.11)$$

The mean variance was used as an assessment for the reproduction variance (3.12). The value of the regression coefficients was verified using the Student's test $H_o : t_i > t_{cr}$ [192]:

$$\sigma_{repr}^2 = \sum \left(\sigma_i^2\right) / n \tag{3.12}$$

$$t_i = |b_i| / \sigma_{b_i} \tag{3.13}$$

where $\sigma_{b_i} = \sqrt{\dfrac{\sigma_{repr}^2}{n \cdot m}}$ – the error of the regression coefficient; n – the number of experiments; m – the number of repetitions of each experiment.

The adequacy of the obtained regression model was verified on the basis of assessing the deviation of the model results from the experimental data on the basis of F – Fisher's test $H_o : F_{on} < F_{cr}$ [192].

$$F_{on} = \sigma_{im}^2 / \sigma_{repr}^2 \tag{3.14}$$

$$\sigma_{im}^2 = \frac{m}{n-1} \sum_{i=1}^{n} (\bar{y}_i - \hat{y}_i)^2 \tag{3.15}$$

where $\hat{y}_i$ – the value of the response function predicted by the regression equation for the *i*-th experiment.

Preliminary processing of the experimental data indicated the homogeneity of the reproducibility variance by G – test, with the degrees of freedom $f_1 = 9$ and $f_2 = 8$ and the level of significance 0.05: $H_o : G = 0.344 \le G_{cr} = 0/3584$.

The regression equation has the form (3.16), the value of the coefficient of determination R^2 indicates the explanation of 99.99% of the variability of the response Y, the assessment of standard error indicates the deviation of 0.09, which gives a satisfactory accuracy of ± 0.03% of the variable component of IM electromagnetic torque:

$$\begin{aligned} y = {} & 6.75 + 2.43X_1 - 0.038X_2 - 0.5X_3 + 0.041X_1X_2 - \\ & - 1.13X_1X_3 - 0.006X_2X_3 + 0.003X_1X_2X_3 \end{aligned} \tag{3.16}$$

The statistical analysis of model (3.16) revealed the significance of the coefficients for factors and combinations of factors X_1, X_2, X_3, X_1X_2 and X_1X_3, (Table 3.7), because the value of the Student's coefficient for them exceeds the tabular value of 1.9934 with the level of significance $\alpha = 0.05$. The response function has taken the form:

$$y = 6.75 + 2.43X_1 - 0.038X_2 - 0.5X_3 + 0.041X_1X_2 - \\ -1.13X_1X_3 \tag{3.17}$$

The obtained model is non-linear. The graph of the response function and the area corresponding to the domain of values of the function $y(X_1, X_2, X_3)$, are shown in Figure 3.31.

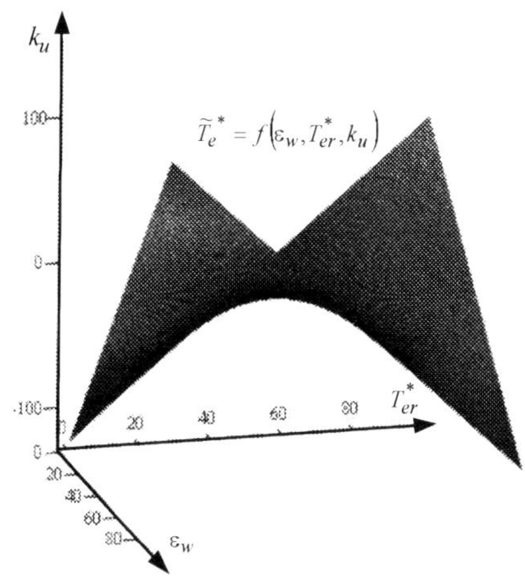

Figure 3.31. The projection of surface $y(X_1, X_2, X_3)$ in the selected area of factor space.

The main results of regression analysis:

- R^2 – the statistics is 99.98%, i.e., more than 99% of the variability of the response is caused by the changes in these factors;
- the corrected value R^2 (taking into account the number of regression variables) is 99.98;
- the standard error of the residuals is 0.009;
- the average value of the residuals is 0.003.

The Darbin-Watson (DW) statistic is 2.5, indicating that there is no significant correlation with the order of the data in the Table. The adequacy of the proposed model based on the assessment of F – statistics is determined, $H_o : F_{on} = 0.099 < F_{cr} = 3.863$ at the power of the criterion 0.95. The analysis of the assessments did not reveal pairwise correlation coefficients between the elements of the model of significant correlation with absolute values exceeding 0.5.

Table 3.8. The assessment of the significance of model coefficients

Parameter	Standard assessment	Error	Student's coefficient	t_{cr} at α=0.05
Constant	6.75167	0.00333333	1734	1.993464
A:X_1	4.855	0.00666667	623.587	1.993464
B:X_2	-0.0766667	0.00666667	9.849	1.993464
C:X_3	-1.00833	0.00666667	129.496	1.993464
AB	0.0816667	0.00666667	10.491	1.993464
AC	-2.25	0.00666667	1.5	1.993464
BC	-0.0116667	0.00666667	288.98	1.993464
ABC	0.00666667	0.00666667	0.857	1.993464

As a result of the experiment, a regression model (3.17) was obtained, which characterizes the relative value of the variable component of the electromagnetic torque based on input factors in a closed SC, and shows the VFED parameters influence on the original coordinate. The statistical adequacy of the model is confirmed by the determination criterion R^2 and the minimum value of standard errors of the equation structural parameters.

As a result of modeling VFED operation the following was obtained: model value $\tilde{T}_e^*$; a standard error for each model value; 95% confidence interval of the model value and 95% confidence interval of the mean model value. The results of modeling are given in Table 3.9.

The analysis of the discrepancies between the data of the regression model and the simulation model in Matlab Simulink according to the initial parameter is presented in Table 3.10.

The statistical analysis of the model (3.17) showed significant reliability of the model relative to the model data. According to the results of the analysis, the greatest influence on the amplitude of the variable component of IM electromagnetic torque is exerted by the first and third factors, that is, the degree of asymmetry of the phases of the motor stator and the gain coefficient of the feedback in the control system, which is confirmed by the results of the preliminary research.

Table 3.9. The expected results of the response of the model function

No.	Model values, %	Lower 95% confident level of model value	Upper 95% confident level of model value	Lower 95% confident level of the mean model value	Upper 95% confident level of the mean model value
1	3.782	3.61263	3.9407	3.66461	3.88872
2	5.024	4.86596	5.19404	4.91794	5.14206
3	3.624	3.46596	3.79404	3.51794	3.74206
4	4.866	4.69596	5.02404	4.74794	4.97206
5	10.806	10.636	10.964	10.6879	10.9121
6	7.547	7.3893	7.71737	7.44128	7.66539
7	10.811	10.6526	10.9807	10.7046	10.9287
8	7.553	7.38263	7.7107	7.43461	7.65872

Table 3.10. The analysis of discrepancies between the data of the regression model and the simulation model

No.	Experiment in Matlab Similink, %	Regression model, %	Residual, %
1	3.77	3.782	-0.0092
2	5.03	5.024	0.0091
3	3.63	3.624	0.0089
4	4.85	4.866	-0.0091
5	10.803	10.806	-0.0025
6	7.55	7.547	0.0027
7	10.813	10.811	0.0025
8	7.55	7.553	-0.0022

The calculated regression model of VFED operation with setting the SC to compensate for the variable component of IM electromagnetic torque makes it possible to solve the problem of analytical calculation of the gain coefficient in the direct channel of the SC under the known load, the degree of phase asymmetry and the desired level of compensation for the variable component of the electromagnetic torque. However, the obtained model (3.17) does not allow solving this problem, because it is nonlinear and takes into account the combined influence of combinations of factors: X_1X_2 and X_1X_3. Therefore, it is necessary to linearize the obtained model and do not take into account the influence of these combinations of factors, which will lead to some deterioration of the reliability of the model.

The linearized response function has taken the form:

$$y = 6.75 + 2.43X_1 - 0.038X_2 - 0.5X_3 \tag{3.18}$$

The obtained model is linear. The graph of the response function and the area corresponding to the domain of the values of function $y(X_1, X_2, X_3)$ are shown in Figure 3.32.

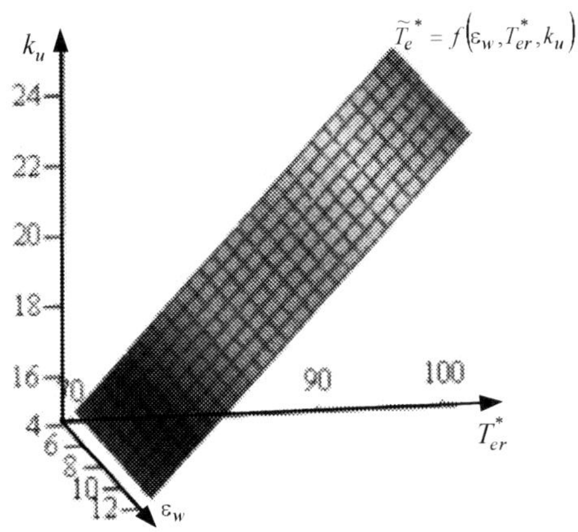

Figure 3.32. The projection of surface $y(X_1,X_2,X_3)$ in the selected area of the factor space.

The results of regression analysis (linearized model):

- R^2 – the statistics is 82.91%, i.e., more than 82% response variability is due to changes in these factors;
- the corrected value R^2 (taking into account the number of variables of the regression) is 70.1%;
- the standard error of residual is 1.59;
- the mean value of residual 1.125.

The statistics of Darbin-Watson (DW) is 2.57, which indicates the absence of significant correlation with the order of data in the Table. The adequacy of the proposed model based on the assessment of F – statistics is determined, $H_o : F_{on} = 2.261 < F_{cr} = 3.482$ with the criterion power 0.95. The analysis of the assessments of pairwise correlation coefficients between elements of the model did not reveal significant correlation with absolute values exceeding 0.5.

As a result of modeling VFED operation the following was obtained: model value $\widetilde{T}_e^*$; the standard error for each model value; 95% confidence interval of the model value and 95% confidence interval of the mean model value. The results of modeling are given in Table 3.11.

Table 3.11. The expected results of the response of the model function

No.	Model value, %	Student residues (model value error)	Lower 95% confident level of model value	Upper 95% confident level of model value	Lower 95% confident level of the mean model value	Upper 95% confident level of the mean model value
1	3.742	-0.962133	-0.547071	10.2804	1.74104	7.99229
2	4.983	1.05964	-1.5554	9.27207	0.732711	6.98396
3	3.665	-1.03712	-0.623737	10.2037	1.66438	7.91562
4	4.907	0.941183	-1.63207	9.1954	0.656044	6.90729
5	10.846	0.948764	4.30793	15.1354	6.59604	12.8473
6	7.588	-1.04527	3.2996	14.1271	5.58771	11.839
7	10.77	1.05141	4.23126	15.0587	6.51938	12.7706
8	7.512	-0.954477	3.22293	14.0504	5.51104	11.7623

The analysis of the discrepancies between the data of the regression model and the simulation model in Matlab Simulink is presented in Table 3.12.

Table 3.12. The analysis of discrepancies between the data of the regression model and the simulation model

No.	Experiment in Matlab Simulink, %	Regression model, %	Residual, %
1	3.77	3.742	0.032
2	5.03	4.983	0.05
3	3.63	3.665	-0.032
4	4.85	4.907	-0.05
5	10.80	10.846	-0.043
6	7.55	7.588	-0.038
7	10.81	10.77	0.043
8	7.55	7.512	0.038

The obtained regression model (3.18) characterizes with sufficient accuracy the relative value of the variable component of the electromagnetic torque based on the input factors in a closed SC. Solving this equation with respect to the third factor, we can find the required value of the gain coefficient in SC at the known parameters X1 and X2 and the desired value of the variable component of IM electromagnetic torque.

$$X_3 = y - \left(\frac{b_0 + b_1 X_1 + b_2 X_2}{b_3} \right) = \\ = y - \left(\frac{6.75 + 2.43 X_1 - 0.038 X_2}{-0.5} \right) \tag{3.19}$$

or

$$k_u = \widetilde{T}_e^{\,*} - \left(\frac{b_0 + b_1 \varepsilon_w + b_2 T_{er}^{*}}{b_3} \right) \tag{3.20}$$

For example, for the modes characterized by the input parameters X_1 and X_2 (Figure 3.33, Table 3.13), corresponding to the degree of asymmetry of the stator phases and the moment of resistance of IM, and to reduce the variable component of IM electromagnetic torque to $\widetilde{T}_e^{\,*} = 2, 4, 7, 10\,\%$, the following value of the gain coefficient is required.

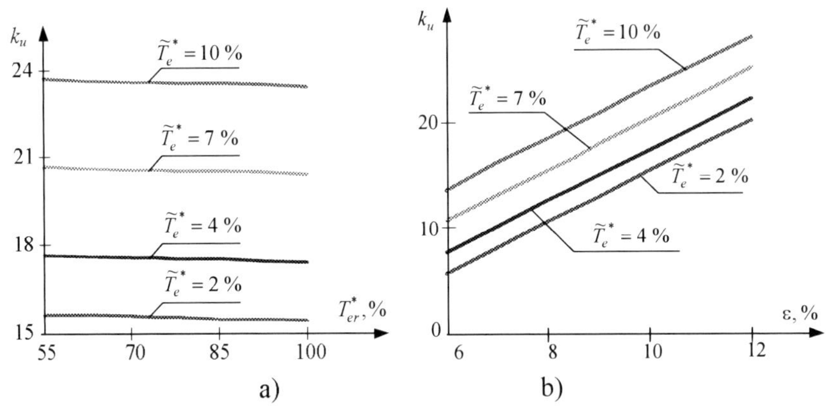

Figure 3.33. The dependence of value k_u: a) on the relative moment of resistance on IM shaft at asymmetry in phase A equal to 10% ($\varepsilon_{wA} = 0.9$); b) on the degree of asymmetry of the stator phases at $T_{er}^{*} = 100\ \%$.

According to the results of calculations and modeling (Table 3.13), the obtained regression model in linearized form allows one to calculate with high accuracy the gain coefficient in SC direct channel, necessary to obtain the desired value of the original coordinate and the analytical calculation of which is quite a difficult task, taking into account the known other factors affecting the amplitude of the variable component of IM electromagnetic torque. It should be noted that the specified value of the accuracy of the obtained results remains within the initial variation of these independent input factors. The obtained dependence for the calculation of the gain coefficient can be applied beyond the described limits, but this will deteriorate SC accuracy.

Table 3.13. The analysis of the discrepancies between the data of the regression model and the simulation model

No.	Value of factor X1		Value of factor X2		Value of factor X3 k_u	Desired value $\tilde{T}_e^*$, %	Value Y (imitation model) $\tilde{T}_e^*$, %	Calculation error Y, %
	Relative	Absolute ε_w, %	Relative	Absolute T_{er}^*, %				
1	0.5	11	0.5	92.5	20	5	5.2	4
2	0.5	11	0	85	21	5	5.2	4
3	0	10	0.5	92.5	18.5	5	4.7	6
4	-0.5	9	-0.5	77.5	15	4	4.3	7

The obtained gain coefficient dependence on the load level and the specified level of compensation for the variable component makes it possible to build VFED adaptive control system (Figure 3.34) [193].

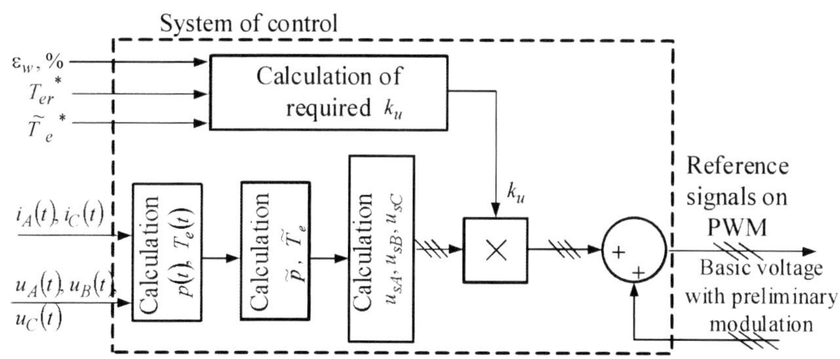

Figure 3.34. The functional diagram of VFED adaptive system of control.

The principle of action of the given SC at compensation for the influence of IM asymmetry consists in the following: on the basis of the signals of instantaneous voltage ($u_A(t), u_B(t), u_C(t)$) and current ($i_A(t), i_C(t)$), measured by means of sensors, the three-phase active power and electromagnetic torque ($p(t), T_e(t)$) are calculated. Variable components ($\tilde{p}, \tilde{T}_e$) are singled out in them. After that, the compensation voltages for each phase (u_{sA}, u_{sB}, u_{sC}) are calculated and fed to SVI control input by means of a proportional voltage regulator. The parameters of this controller depend on the degree of IM asymmetry, the level of its load and the set level of compensation for the variable component of the original coordinate.

During the compensation for the variable component of the electromagnetic torque, value k_u is calculated as follows:

1. The stator winding asymmetry degree ε_w is determined in the mode of the pre-start identification of IM parameters, the desired value of the electromagnetic torque variable components $\tilde{T}_e^*$ is set as minimal as possible (1–2)% as the PWM supply voltage $\tilde{T}_e^* = 0\ \%$ is practically unattainable), the value of static resistance moment T_{er}^* the current mode of operation in accordance with the method of indirect determination of IM energy performance.
2. Based on the set $\tilde{T}_e^*$ and determined parameters ε_w and T_{er}^* linearized function $k_u = f\left(T_{er}^*\right)$ is created for the determination of the regulator coefficient for each particular IM. In order to find the reference points of this function in industrial conditions, it is necessary to build two dependencies $\tilde{T}_e^* = f(k_u)$ with two basic values T_{er}^* four test runs of ED in the pre-start adjustment mode, we get two basic values k_u, on the basis of which the linear dependence $k_u = f\left(T_{er}^*\right)$ is built. When compensating for the variable component of power, this dependence is determined similarly.

3.5. The Calculation of the Recommended Loading Level of an Asymmetric Induction Motor

The proposed system of compensation for variable components of power consumption and electromagnetic torque of an induction motor can significantly reduce the negative impact of IM asymmetry, which caused the appearance of these components, on its mode of operation and its residual life.

However, such an approach cannot always be implemented on a specific industrial site, for example, due to the lack of such control or power equipment, which is necessary for the implementation of the above control algorithms. It results in the fact that the motor, even with a slight asymmetry of parameters operates in a mode characterized by significant asymmetry of currents, and, consequently, the uneven distribution of losses in the windings. Such operation causes a further deterioration of the technical condition of the electric machine, which occurs due to the increase in the level of parameter asymmetry, the level of excess losses. Thus, without correcting the mode of operation of asymmetric IM, the consequence of the combined effect of the described factors will consist in a significant reduction in IM residual life, which is determined by the state of the windings insulation [183, 184].

Taking this into consideration, one of the ways to increase the efficiency of ED with asymmetric IM and decrease the rate of reduction of its service life is to reduce the maximum allowable load [194], which can be determined from the static characteristics of IM in its asymmetry.

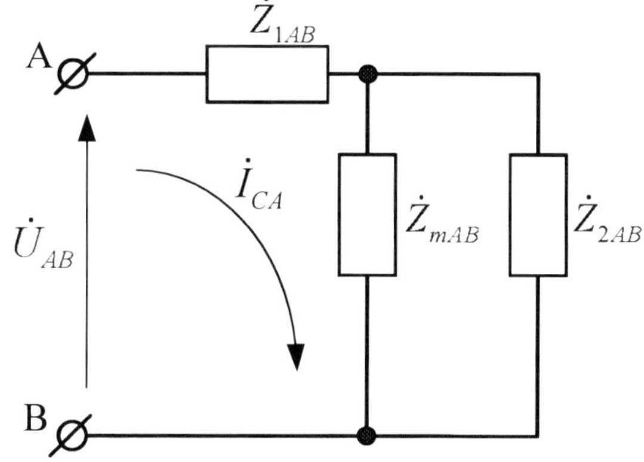

Figure 3.35. T-shaped equivalent circuit of the two phases of IM stator when connecting the windings according to the scheme "star without zero."

To research the IM asymmetry influence on its static characteristics, it is necessary to take into account the fact that a typical algorithm for constructing static characteristics on a T-shaped or L-shaped EC cannot be used because it assumes that the induction motor is symmetrical [195-197]. Therefore, the paper proposes the following sequence of calculation of IM static characteristics, taking into account the asymmetry of its parameters.

In the first stage, a T-shaped equivalent circuit is formed for each phase of the motor. Figure 3.35 presents an EC, which describes the circuit of the current flux between phases A-B, EC for the circuits B-C and C-A are composed similarly [198].

The following designations are accepted on the equivalent circuit:

- $\dot{Z}_{1AB}$ – total complex resistance of the stator phases A and B;
- $\dot{Z}_{mAB}$ – total complex resistance of the magnetizing circuit between the stator phases A and B;
- $\dot{Z}_{2ab}$ – total complex resistance of the rotor phases a and b.

As an example, calculations were performed for the case of asymmetry of phase A. Taking this into account, EC total complex resistances are determined as follows [198, 199].

The stator circuit:

$$\begin{aligned}
\dot{Z}_{1AB} &= \left[R_A(1-\varepsilon_w)+R_B\right]+j\left[\omega(L_A(1-\varepsilon_w)^2)+\omega L_B\right];\\
\dot{Z}_{1BC} &= \left[R_B+R_c\right]+j\left[\omega L_B+\omega L_C\right];\\
\dot{Z}_{1CA} &= \left[R_C+R_A(1-\varepsilon_w)\right]+j\left[\omega L_C+\omega(L_A(1-\varepsilon_w)^2)\right],
\end{aligned} \tag{3.21}$$

where R_A, R_B, R_C – the stator phases resistances; L_A, L_B, L_C – stator phase scattering inductance; ε_w – asymmetry coefficient; ω – angular frequency of the power supply network.

Magnetizing circuits:

$$\begin{aligned}
\dot{Z}_{mAB} &= j\left[\omega L_{mA}(1-\varepsilon_w)+\omega L_{mB}\right];\\
\dot{Z}_{mBC} &= j\left[\omega L_{mB}+\omega L_{mC}\right];\\
\dot{Z}_{mCA} &= j\left[\omega L_{mC}+\omega L_{mA}(1-\varepsilon_w)\right],
\end{aligned} \tag{3.22}$$

where R_m – the resistance of magnetizing circuit; $L_A = L_B = L_C = L_m$ – the inductance of the magnetizing circuit.

The rotor circuit:

$$\dot{Z}_{2ab} = \frac{R_a + R_b}{s} + j(\omega L_a + \omega L_b);$$
$$\dot{Z}_{2bc} = \frac{R_b + R_c}{s} + j(\omega L_b + \omega L_c);$$
$$\dot{Z}_{2ca} = \frac{R_c + R_a}{s} + j(\omega L_c + \omega L_a), \tag{3.23}$$

where R_a, R_b, R_c – the rotor phases resistances; L_a, L_b, L_c – rotor phase scattering inductance, s – slip.

To determine the currents in the circuits it is necessary to determine the total resistance at the terminals A and B of the stator windings, for this the total resistance of the parallel branches of the magnetization circuits and the rotor is calculated first:

$$\dot{Z}_{m2ab} = \frac{\dot{Z}_{mAB}\dot{Z}_{2ab}}{\dot{Z}_{mAB} + \dot{Z}_{2ab}}; \dot{Z}_{m2bc} = \frac{\dot{Z}_{mBC}\dot{Z}_{2bc}}{\dot{Z}_{mBC} + \dot{Z}_{2bc}};$$
$$\dot{Z}_{m2ca} = \frac{\dot{Z}_{mCA}\dot{Z}_{2ca}}{\dot{Z}_{mCA} + \dot{Z}_{2ca}}. \tag{3.24}$$

General total complex resistance:

$$\dot{Z}_{AB} = \dot{Z}_{1AB} + \dot{Z}_{m2ab}; \dot{Z}_{BC} = \dot{Z}_{1BC} + \dot{Z}_{m2bc};$$
$$\dot{Z}_{CA} = \dot{Z}_{1CA} + \dot{Z}_{m2ca}. \tag{3.25}$$

Linear currents flowing in the circuits A-B, B-C, C-A are determined by Ohm's law:

$$\dot{I}_{AB} = \frac{\dot{U}_{AB}}{\dot{Z}_{AB}}; \dot{I}_{BC} = \frac{\dot{U}_{BC}}{\dot{Z}_{BC}}; \dot{I}_{CA} = \frac{\dot{U}_{CA}}{\dot{Z}_{CA}}, \tag{3.26}$$

where $\dot{U}_{AB} = \sqrt{3}U_{\max} e^{j\frac{\pi}{6}}$, $\dot{U}_{BC} = U_{BC} e^{j\frac{\pi}{3}}$, $\dot{U}_{CA} = U_{CA} e^{j\frac{\pi}{3}}$ – the linear voltages in a complex form.

Then, according to Kirchhoff's law, the currents in the stator phases are determined by expression:

$$\dot{I}_A = \dot{I}_{AB} + \dot{I}_{CA}; \dot{I}_B = \dot{I}_{BC} + \dot{I}_{AB}; \dot{I}_C = \dot{I}_{CA} + \dot{I}_{BC} \tag{3.27}$$

The interphase value of electromotive force:

$$\dot{E}_{AB} = \dot{U}_{AB} - \dot{I}_{AB} \cdot \dot{Z}_{1AB}; \dot{E}_{BC} = \dot{U}_{BC} - \dot{I}_{BC} \cdot \dot{Z}_{1BC};$$
$$\dot{E}_{CA} = \dot{U}_{CA} - \dot{I}_{CA} \cdot \dot{Z}_{1CA}. \tag{3.28}$$

Linear rotor currents are determined similarly to linear stator currents:

$$\dot{I}_{2ab} = \frac{\dot{E}_{AB}}{\dot{Z}_{m2ab} + \dot{Z}_{2ab}}; \dot{I}_{2bc} = \frac{\dot{E}_{BC}}{\dot{Z}_{m2bc} + \dot{Z}_{2bc}};$$
$$\dot{I}_{2ca} = \frac{\dot{E}_{CA}}{\dot{Z}_{m2ca} + \dot{Z}_{2ca}}. \tag{3.29}$$

The currents in the rotor phases:

$$\dot{I}_{2a} = \dot{I}_{2ab} + \dot{I}_{2ca}; \dot{I}_{2b} = \dot{I}_{2bc} + \dot{I}_{2ab}; \dot{I}_{2c} = \dot{I}_{2ca} + \dot{I}_{2bc}. \tag{3.30}$$

It is known that in asymmetric modes, the total electromagnetic torque of the motor consists of two components: moments of forward and reverse sequence determined by the corresponding currents of the rotor. Then, the rotor currents of the forward and reverse sequence are as follows:

$$\dot{I}_{2dir} = \frac{1}{3}\left(\dot{I}_{2a} + \dot{I}_{2b}a + \dot{I}_{2c}a^2\right) \tag{3.31}$$

$$\dot{I}_{2rev} = \frac{1}{3}\left(\dot{I}_{2a} + \dot{I}_{2b}a^2 + \dot{I}_{2c}a\right) \tag{3.32}$$

Thus, the electromagnetic torque of the forward and reverse sequence:

$$T_{e\ dir} = \frac{3\left|\dot{I}_{2dir}\right|^2 R_2}{\omega_0 s} \quad (3.33)$$

$$T_{e_{rev}} = \frac{3\left|\dot{I}_{2rev}\right|^2 R_2}{-\omega_0 (2-s)}. \quad (3.34)$$

The total electromagnetic torque of the motor:

$$T_e = T_{e\ dir} + T_{e_{rev}}. \quad (3.35)$$

Thus, the presented IM model makes it possible to calculate and analyze the static characteristics of the motor taking into account its asymmetry. In addition, in contrast to the model for the research of dynamic characteristics, this model requires much smaller calculation operations.

On the basis of the given technique of construction of IM mechanical characteristic at the asymmetry of its parameters (3.20-3.35) it is possible to calculate the maximum admissible level of IM loading at various degrees of asymmetry taking into account the assumption that the level of allowable load in the steady state is determined by electrical losses, which should be equal to the losses in the symmetrical mode. It is also assumed that changes in steel losses can be neglected and admitted to remain unchanged. However, at the same time the change of sliding at IM asymmetry should be considered.

Figures 3.36-3.39 present the results of calculation of currents and mechanical characteristics for the case of 20% stator asymmetry:

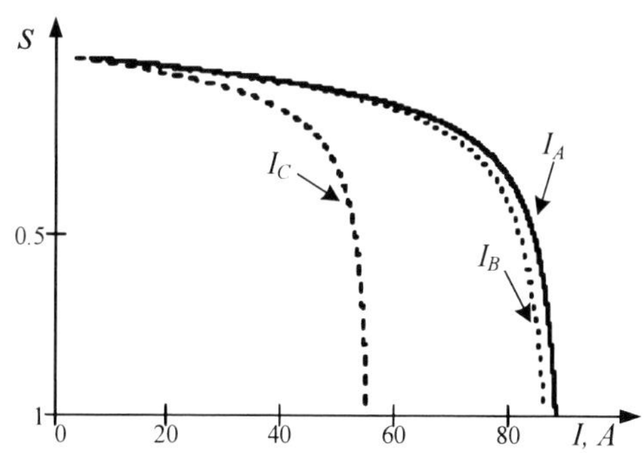

Figure 3.36. IM stator currents with asymmetry in phase A equal to 20% ($\varepsilon_{wA} = 0.8$).

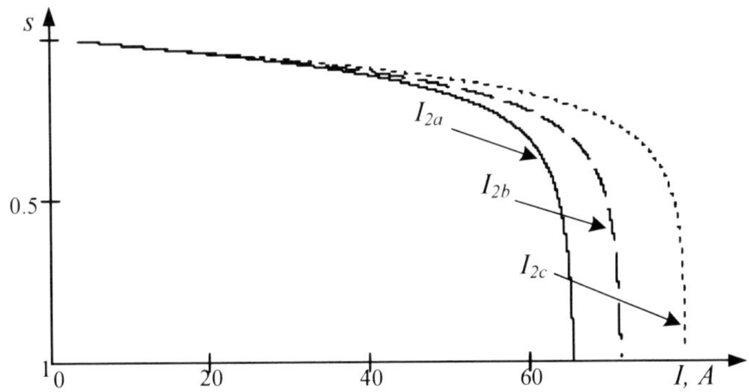

Figure 3.37. IM rotor currents with asymmetry in phase A equal to 20% ($\varepsilon_{wA} = 0.8$).

Figure 3.39 shows IM mechanical characteristics taking into account the moments of forward and reverse sequence with 20% asymmetry of the stator phases. The analysis of the constructed dependences indicates that at this level of asymmetry the moment of the reverse sequence is significant, which, in turn, results in a decrease in the value of the starting and critical moments on the total mechanical characteristics.

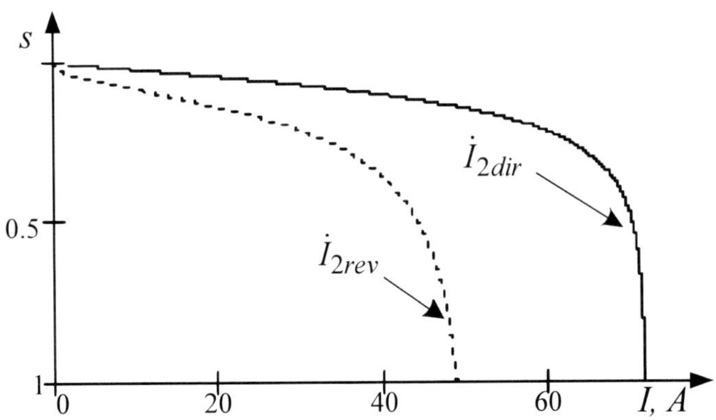

Figure 3.38. The rotor currents of the forward and reverse sequences with asymmetry in phase A equal 20% ($\varepsilon_{wA} = 0.8$).

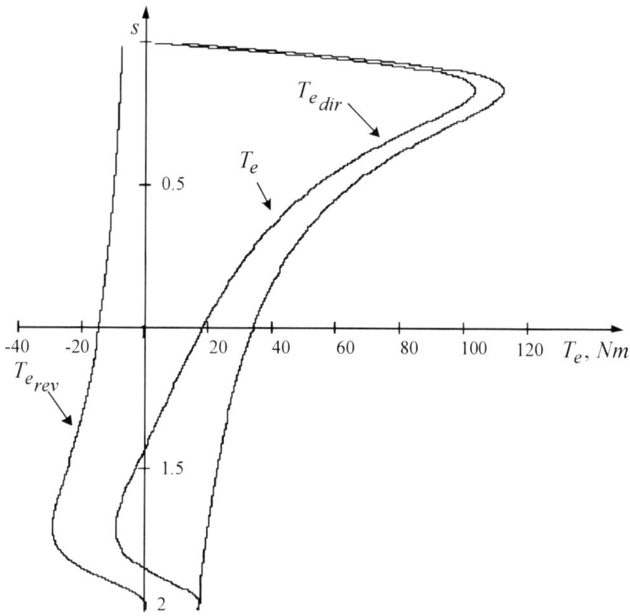

Figure 3.39. Static mechanical characteristics of IM and moments of forward and reverse sequence moment with asymmetry in phase A equal to 20% ($\varepsilon_{wA} = 0.8$).

The mechanism for calculating the allowable maximum load for each degree of asymmetry ε is as follows:

1. Slip s is determined on the constructed mechanical characteristic, which corresponds to the rated load.
2. The stator phase currents ($I_A/I_B/I_C$) are determined by this slip.
3. The total and phase losses in the stator copper are calculated at the determined currents ($\Delta P_{Cu1}/\Delta P_{CuA}/\Delta P_{CuB}/\Delta P_{CuC}$).
4. Such value of sliding at which currents in stator phases will cause the electric losses equal to the rated ones is determined on the current characteristic corresponding to the set degree of asymmetry.
5. The maximum allowable load moment ($M^{/}$) is determined at the calculated slip value.

Table 3.14 contains the results of calculating the maximum allowable load in terms of allowable winding temperature for stator phase asymmetry of 5, 10, 15 and 20%. The calculation of the recommended load level was performed for two cases:

- meeting the condition of equality of total electrical losses in the stator with their rated value (M'_1);
- meeting the condition of equality of electrical losses in the most superheated phase winding with their rated value (M'_f).

Table 3.14. The calculation of the recommended level of IM loading at different degrees of asymmetry

Asymmetry level ε_w, %	**Slip s at $T_e = T_{er}$**	**$I_A/I_B/I_C$, A**	**$\Delta P_{Cu1}/\Delta P_{CuA}/\Delta P_{CuB}/\Delta P_{CuC}$, W**	**M'_1/M'_ϕ, %**
0%	0.028	11.7	303/101/101/101	100%
5%	0.028	12.7/12.6/11.7	328/110/116/101	94%/92%
10%	0.029	14.4/14.1/12.2	395/136/151/108	87%/80%
15%	0.031	17.6/17.5/12.8	534/189/224/121	72%/64%
20%	0.0348	22.4/22.5/14.3	810/293/368/149	52%/41%

The calculation data show that even at 20% phase asymmetry electrical losses in the stator increase almost by 2.5 times, which reduces the recommended load level to 52%, if assessed by the total electrical losses in the stator, or 41%, if judged by the losses in the most overheated winding. Moreover, the second approach is more correct, because significant overheating of one of the phases, even with a small increase in total losses causes a significant deterioration of the technical condition of the motor.

Conclusion

1. The method and mathematical provision of the method of compensation for the influence of IM asymmetry by means of the variable-frequency electric drive are offered and proved on the basis of the cross-vector theory of instantaneous power. Unlike other theories, it does not require the intermediate transformation of coordinates, which in turn, greatly simplifies the calculation apparatus of the control system and increases its speed.
2. The conformity of the structural model of the variable-frequency electric drive with asymmetric IM with the real modes of control and functioning of measuring and power converter equipment is proved by means of comparative analysis of the results of structural modeling and modeling on the basis of virtual blocks.

3. The results of mathematical modeling prove the effectiveness of the proposed method of compensation, it is shown that the inclusion of the compensator reduces the undesirable variable component of the torque almost by an order of the value, which results in reduced vibration and improved operating conditions in general. The use of the cross-vector theory of instantaneous power allows solving the problem of compensation for both inactive components of three-phase IM power and electromagnetic torque. In the former case, the best results are observed in terms of compensation for the variable component of active power, and in the latter case – according to the electromagnetic torque.
4. It is proved that the adjustment of the control system of the variable-frequency electric drive when compensating for the elimination of the variable component of the consumed three-phase power with asymmetry of windings in the range of 5–15%, allows reducing the thermal overload of individual windings by more than 50%. The latter, in turn, increases the service life of the motor insulation by almost 6 times compared with the operation without compensating for the asymmetry of the windings. It is shown that from the point of view of reduction of overheating of separate phases the operation of ED with adjustment of the control system for the compensation for the variable components of three-phase active power shows the best results, but at the same time the level of compensation for a variable component of electromagnetic torque decreases.
5. It is shown that due to uneven current loading the asymmetry of the parameters of IM stator windings negatively affects the mode of operation of the corresponding power semiconductor switches of the autonomous voltage inverter. It is proved that the application of the developed compensation system improves the operating conditions of SVI power switches and prolongs their service life in case of asymmetry of IM stator windings due to current redistribution and, as a consequence, significantly reduces losses in the most overloaded valves.
6. It is shown that the level of compensation for the variable power component and IM electromagnetic torque depends on the value of the gain coefficient in SC direct channel and the level of motor load. The gain coefficient dependence on the load level obtained by building a regression model and the specified level of compensation

for the variable component, obtained by the creation of a regression model, allows building VFED adaptive control system.

7. An algorithm for calculating the maximum recommended load level of an induction motor is developed for different degrees of asymmetry in terms of ensuring the allowable insulation temperature of the windings when compensation is impossible. The application of this approach in terms of the technological process, which makes it possible to maintain IM load at a certain level, significantly improves the operating conditions of the motor and increases its service life.

Chapter 4

The Correction of the Operation Modes of Induction Motors with Stator Asymmetrical Windings during Vector Control

4.1. The Compensation for the Variable Component of the Electromagnetic Torque of an Induction Motor

In IM operation with asymmetric stator windings, the electromagnetic torque of the motor can be represented by the sum of constant and variable components [122]: $T_e = T_{e0} + \widetilde{T}_e$. Accordingly, to reduce the effect of asymmetry of IM stator windings on the characteristics of the system of a variable-frequency ED with vector control, a method is proposed to compensate for the variable component of the electromagnetic torque of the motor.

According to the proposed method, the value of the variable component of the electromagnetic torque of the motor is singled out in the control system, based on the expression:

$$\widetilde{T}_e = T_e - \frac{1}{T}\int_0^T T_e dt, \tag{4.1}$$

where T_e – the electromagnetic torque of IM; T – is the signal period of the variable component of electromagnetic torque, which is determined at the frequency of the power supply ($T = 1/f_c$).

Taking into account that IM electromagnetic moment is difficult to directly measure, its determination can be provided by an indirect method based on the known values of the torque-generating component of the stator current vector and the rotor flux linkage module pre-calculated in the control system, according to expression [137]:

$$T_e = \frac{3}{2} p_n \frac{L_\mu}{L_{r\sigma} + L_\mu} |\Psi_r| I_{sq}, \tag{4.2}$$

where p_n – the number of poles pairs; Ψ_r – the rotor flux; I_{sq} – the stator q axes current.

The obtained signal of the variable component of the electromagnetic torque $\widetilde{T}_e$

component of the stator current I_q in the channel of speed regulation. Accordingly, the reference signal of IM stator current along axis q is defined as:

$$I'_{q(ref)} = I_{q(ref)} - k_i \widetilde{T}_e, \tag{4.3}$$

where $I_{sq(ref)}$ – the initial signal setting the IM stator current along axis q; k_i – reduction coefficient ($k_i = I_r / T_e$).

The functional system of the proposed vector control system with the function of compensation for the variable component of IM electromagnetic moment is shown in Figure 4.1.

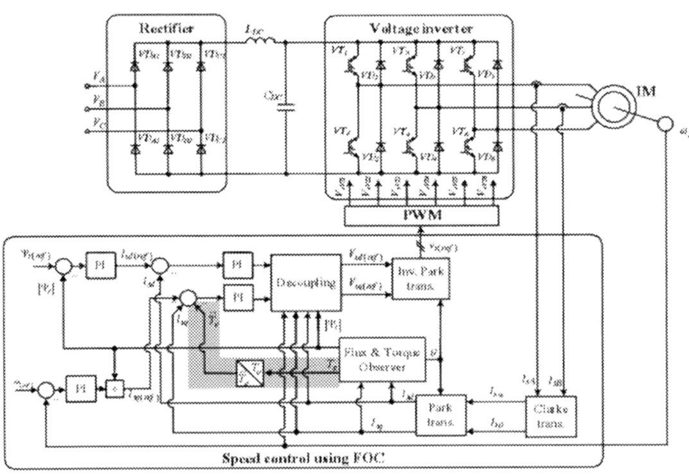

Figure 4.1. The functional diagram of variable-frequency ED with the function of compensation for the variable component of the electromagnetic torque.

The research of operating modes of IM series 4A112M4 with a power of 5.5 kW was carried out on the basis of the developed control system for the following cases: the asymmetry in phase A is 5% ($\varepsilon_{wA} = 0.95$) (*mode No. 1*); the asymmetry in phase A is 10% ($\varepsilon_{wA} = 0.9$) (*mode No. 2*); the asymmetry in phase A is 5%, in phase C – 3% ($\varepsilon_{wA} = 0.95$, $\varepsilon_{wC} = 0.97$) (*mode No. 3*); the asymmetry in phase A is 10%, in phase C – 7% ($\varepsilon_{wA} = 0.9$, $\varepsilon_{wC} = 0.93$) (*mode No. 4*).

The graphs of transients on the electromagnetic torque of the researched IM with asymmetric stator windings in the classical and the proposed control system are presented in Figure 4.2, and their spectral composition is shown in Figure 4.3. Electromagnetic torque signals are given for the 4th operating mode.

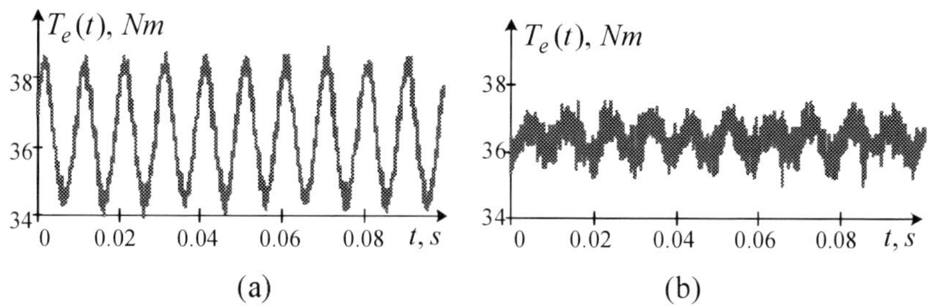

Figure 4.2. The electromagnetic torque of asymmetric IM in the classical (a) and in the proposed (b) control system.

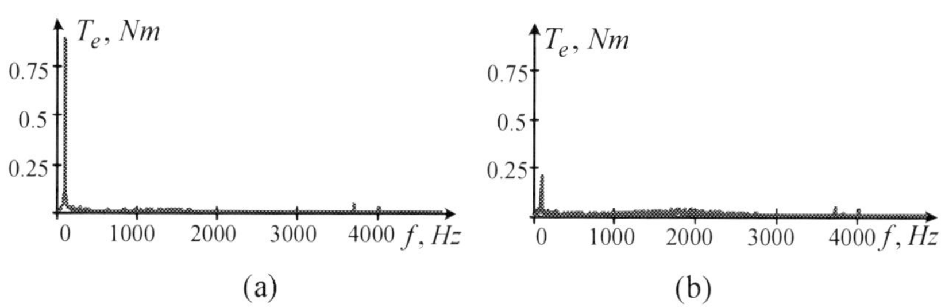

Figure 4.3. The spectral composition of the electromagnetic torque of IM in the classical (a) and in the proposed (b) control system.

The performed research reveals that the application of the proposed control system can reduce the variable component of the electromagnetic torque. Thus, a comparison of their RMS values for these cases of asymmetry of the stator windings of the motor is given in Figure 4.4, where the following notation of the researched parameters is accepted: ▢ – before compensation; ■ – after compensation.

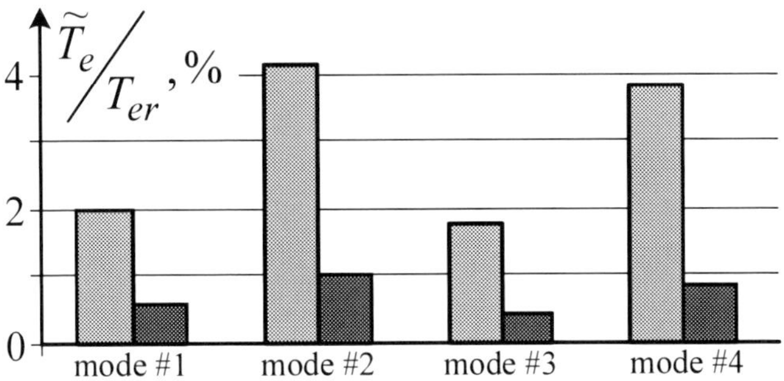

Figure 4.4. The variable components of IM electromagnetic moment.

Thus, the use of the proposed method to compensate for the influence of the asymmetry of the IM windings on the characteristics of the ED system with vector control reduces the variable component of the electromagnetic torque by an average of 73% in asymmetry in one phase of IM and 75% in asymmetry in two phases.

The signals of the active power consumed from the network and their spectral composition for the 4th mode of operation are given in Figures 4.5–4.6.

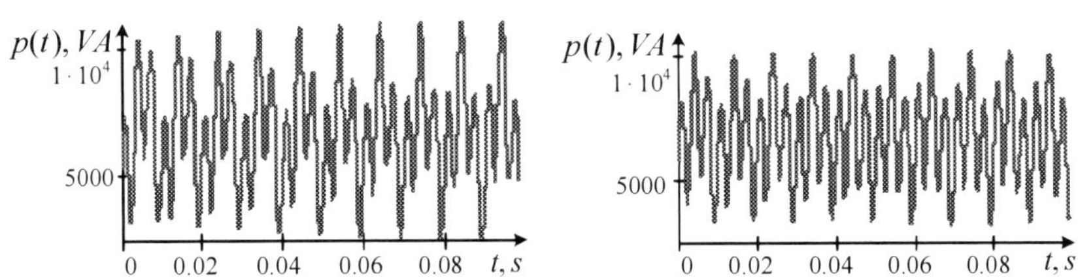

(a) (b)

Figure 4.5. The active power consumed from the network by asymmetric IM in the classical (a) and in the proposed (b) control system.

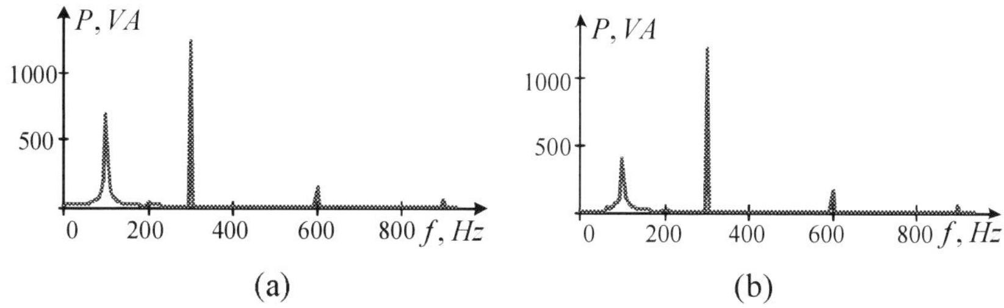

Figure 4.6. The spectral composition of the active power consumed from the network by asymmetric IM in the classical (a) and in the proposed (b) control system.

The research results demonstrate that when using the proposed control system the variable component of the active power consumed from the network also reduces, which can be explained by the balancing of the network current by phases (Figure 4.7).

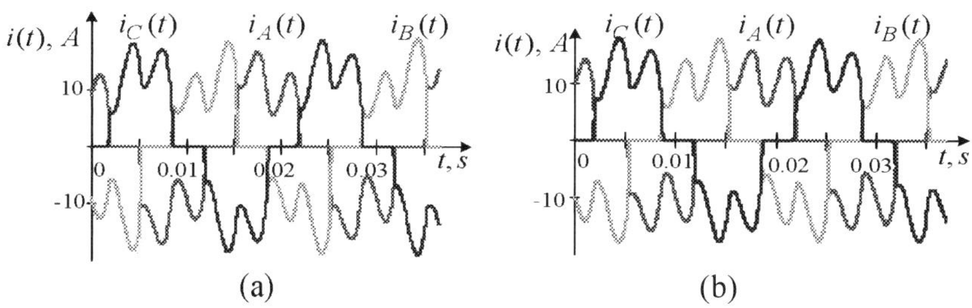

Figure 4.7. The power supply current signals during operation of asymmetric IM in classical (a) and in the proposed (b) control system.

The signals of the active power consumed by FC and their spectral composition for the 4th mode of operation are given in Figures 4.8-4.9.

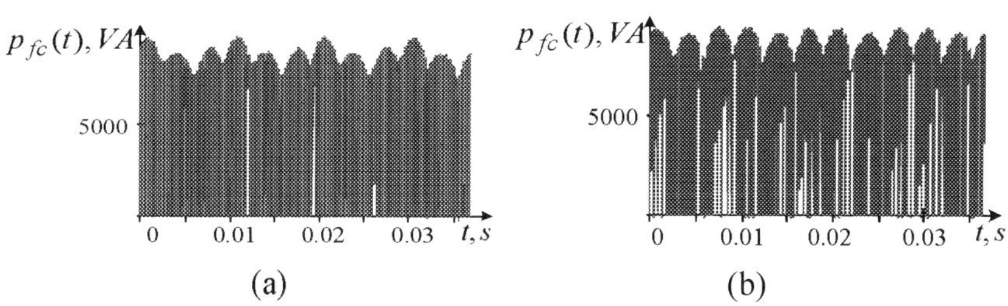

Figure 4.8. The active power of the asymmetric IM consumed by FC in the classical (a) and in the proposed (b) control system.

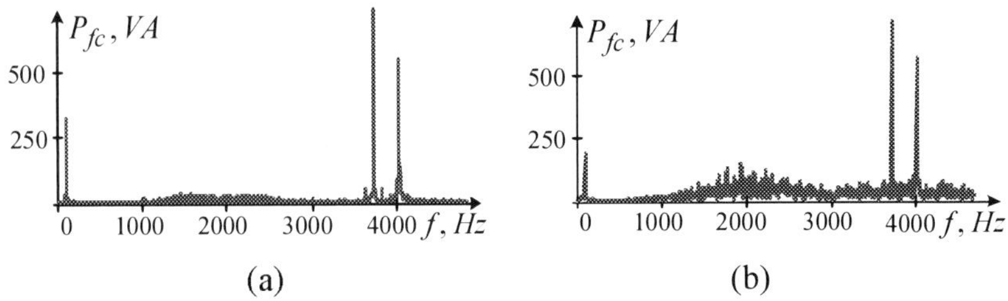

Figure 4.9. The spectral composition of the active power consumed from FC by an asymmetric IM in the classical (a) and in the proposed (b) control system.

Thus, the application of the proposed control system also reduces the variable component of the active power consumed from FC. So, the stator current signals of the researched motor are given in Figure 4.10.

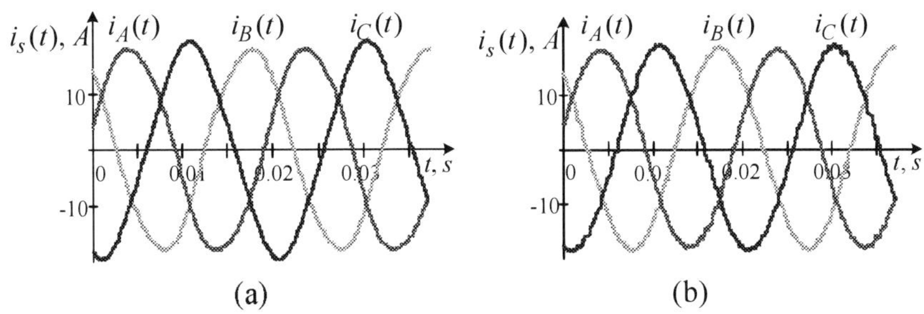

Figure 4.10. The stator current of asymmetric IM in classical (a) and in the proposed (b) control system.

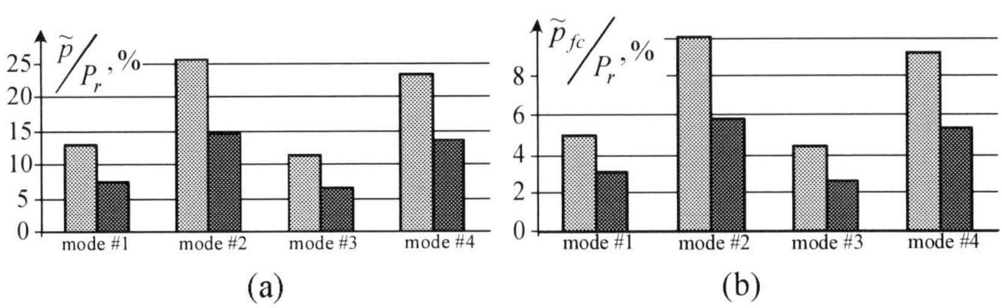

Figure 4.11. The variable components of active power consumed from the network (a) and from FC (b).

A comparison of the root mean square values of the variable components of the power consumed from the network and from FC for these cases of asymmetry of the motor stator windings is given in Figure 4.11.

The results of the research showed that the use of the proposed method of compensation for the asymmetry of IM windings reduces the variable component of active power consumption by 42% on average, and the variable component of power consumed from FC – by 41% with asymmetry in one or two IM phases.

As shown in [200], the asymmetry of IM stator windings can lead to significant overheating of a single winding, even with a slight increase in total losses. In turn, the asymmetric current loading of the motor phases leads to local overheating in some semiconductor switches of the uncontrolled rectifier and autonomous voltage inverter. Therefore, when researching the modes of operation of the compensation system, it is necessary to control not only the variable components of the electromagnetic torque or power consumption, but also the heating losses in IM phases and FC power semiconductor switches. Accordingly, the relative values of the change in losses in the power part of the variable-frequency ED can be calculated based on expression 3.6.

The redistribution of losses in the stator copper by IM phases and in the power elements of FC with these cases of asymmetry of the windings before and after compensation is shown in Figures 4.12-4.13.

The results of the research revealed that with the use of the proposed method of correction of asymmetric modes of variable-frequency ED with vector control, the deviation of losses in the stator copper in the busiest phase can be reduced by an average of 38% in asymmetry in one phase and by 27% in asymmetry in two phases. In turn, the losses in the semiconductor switches of the most loaded phases are also reduced: in the power diodes of the uncontrolled rectifier by an average of 46% in asymmetry in one phase, and by 25% in asymmetry in two phases; in transistor switches by 29% in asymmetry in one phase, and by 30% in asymmetry in two phases; in reverse diodes of power transistors by 9% in asymmetry in one phase, and by 12% in asymmetry in two phases.

Thus, the proposed method of compensating for the asymmetry of IM windings makes it possible to reduce the variable component of the electromagnetic torque of the motor to an acceptable level, with a slight reduction in variable components of power consumption and losses in the power part of variable-frequency ED.

1st mode (asymmetry in phase A is 5%)

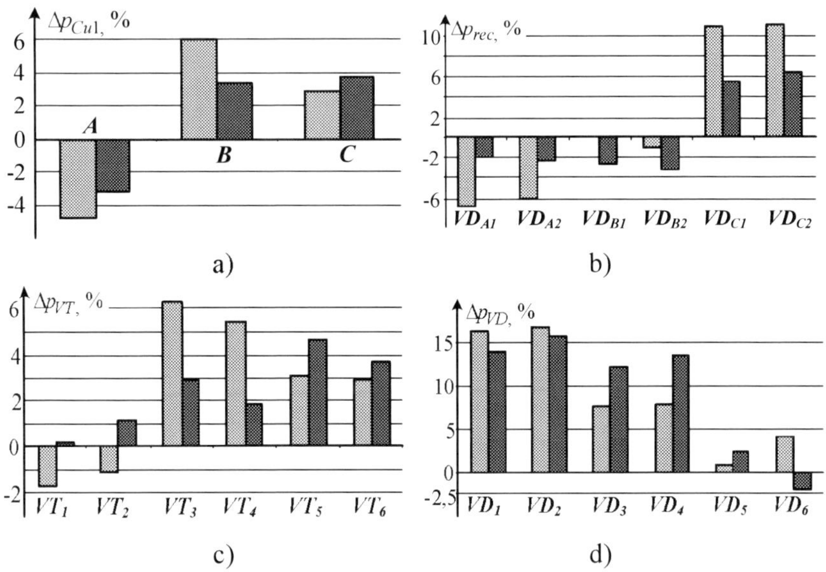

2nd mode (asymmetry in phase A is 10%)

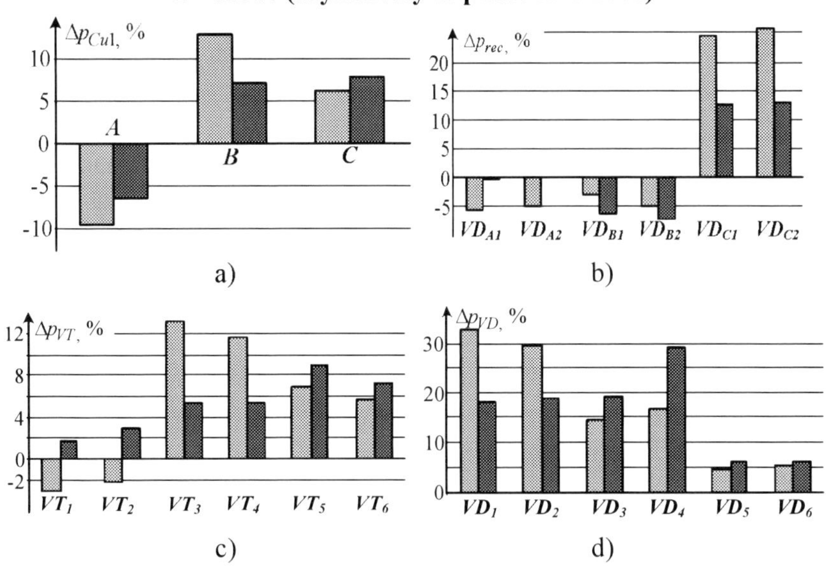

Figure 4.12. The deviation of losses in stator windings (a), input rectifier diodes (b), autonomous inverter transistors (c) and inverse diodes of transistors (d) from their rated values before and after compensation.

3rd mode (asymmetry in phase A is 5%, in phase C – 3%)

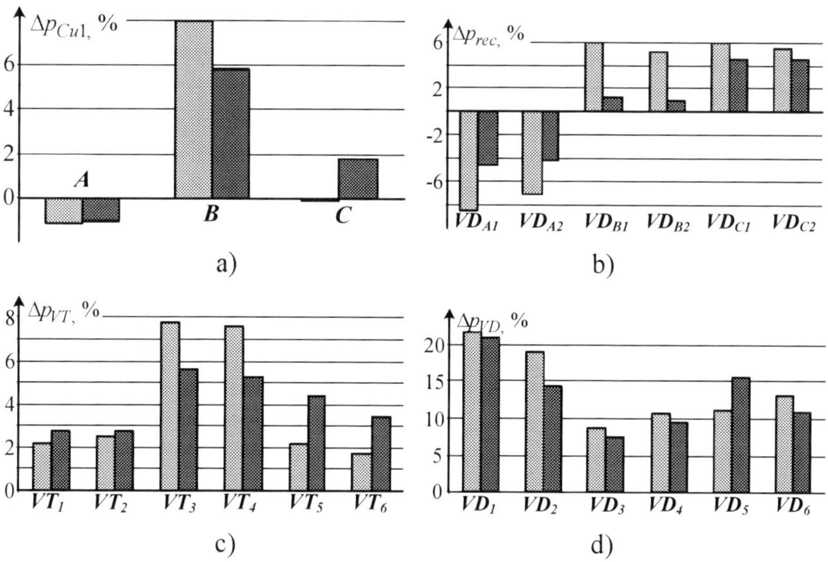

4th mode (asymmetry in phase A is 10%, in phase C – 7%)

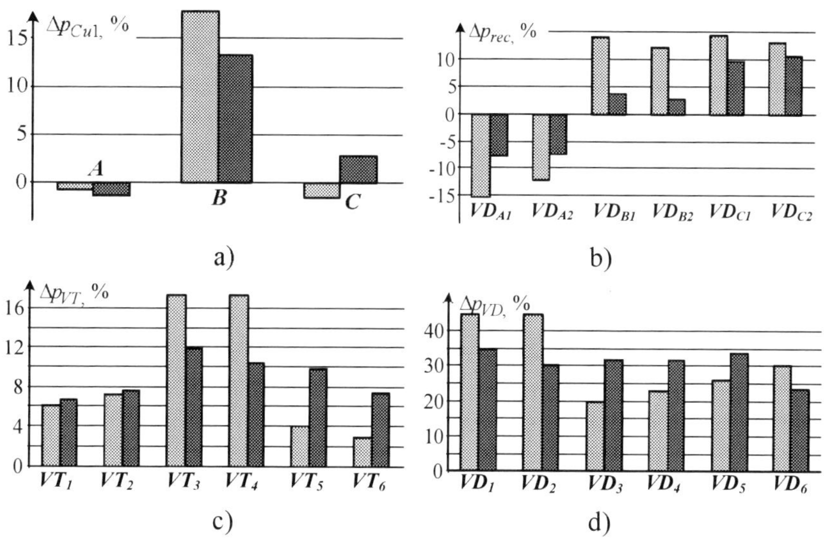

Figure 4.13. The deviation of losses in stator windings (a), input rectifier diodes (b), autonomous inverter transistors (c) and inverse diodes of transistors (d) from their rated values before and after compensation.

4.2. The Compensation for the Variable Component of the Consumed Active Power of the Induction Motor

The classical vector control system provides feedback on the projections of the stator current on the axis of the rotating coordinate system *d-q* [8]. Therefore, one of the known theories of power can be used to compensate for the effect of asymmetry of motor windings on the energy performance of the variable-frequency ED system. Thus, [201] proposes a modified method of *p-q* power theory. It is based on the use of Park and Clarke transformations, and is called the i_d-i_q method.

According to the *p-q* power theory, the conversion of instantaneous values of voltages and currents from the three-phase coordinate system *A-B-C* to the fixed coordinate system α-β-0 is performed on the basis of expressions [202, 203]:

$$\begin{bmatrix} u_\alpha \\ u_\beta \\ u_0 \end{bmatrix} = \sqrt{\frac{2}{3}} \begin{bmatrix} 1 & -1/2 & -1/2 \\ 0 & \sqrt{3}/2 & -\sqrt{3}/2 \\ 1/\sqrt{2} & 1/\sqrt{2} & 1/\sqrt{2} \end{bmatrix} \cdot \begin{bmatrix} u_A \\ u_B \\ u_C \end{bmatrix},$$

$$\begin{bmatrix} i_\alpha \\ i_\beta \\ i_0 \end{bmatrix} = \sqrt{\frac{2}{3}} \begin{bmatrix} 1 & -1/2 & -1/2 \\ 0 & \sqrt{3}/2 & -\sqrt{3}/2 \\ 1/\sqrt{2} & 1/\sqrt{2} & 1/\sqrt{2} \end{bmatrix} \cdot \begin{bmatrix} i_A \\ i_B \\ i_C \end{bmatrix}.$$

Transformation *α-β-0* singles out the component of the zero sequence and the components of the spatial vectors of current and voltage. The component on axis *0* corresponds to the zero sequence, and the component on axes α-β corresponds to the forward and reverse rotational sequences. Active instantaneous power $p_{\alpha\beta}$ and reactive instantaneous power $q_{\alpha\beta}$ in area α-β is determined as the scalar and vector product of voltage and current vectors, respectively:

$$\begin{cases} p_{\alpha\beta} = u_\alpha i_\alpha + u_\beta i_\beta; \\ q_{\alpha\beta} = u_\alpha i_\beta - u_\beta i_\alpha. \end{cases}$$

In the case of stator asymmetry, the instantaneous active and reactive power can be decomposed into two components – constant and variable:

$$\begin{cases} p(t) = P_0 + \tilde{p}(t); \\ q(t) = Q_0 + \tilde{q}(t). \end{cases}$$

In a three-phase network in steady operation, the constant components of active P_0 Q_0 power are determined by the fundamental harmonic of the direct sequence network current, and variable components $\tilde{p}(t)$ and $\tilde{q}(t)$ are determined by load current harmonics other than the main one, and the current of the main harmonics of the reverse sequence.

In the general case, to compensate for the variable components of the active and reactive power, the setting of the compensation current is based on expressions:

$$\begin{bmatrix} I_{c\alpha}^{*} \\ I_{c\beta}^{*} \end{bmatrix} = \frac{1}{\sqrt{U_\alpha^2 + U_\beta^2}} \begin{bmatrix} U_\alpha & -U_\beta \\ U_\beta & U_\alpha \end{bmatrix} \begin{bmatrix} -\tilde{p} \\ -\tilde{q} \end{bmatrix}. \tag{4.4}$$

Since the vector control system is synthesized in the rotating coordinate system, according to the id-iq method, the transition from signals of phase currents and voltages in the fixed coordinate system to the rotating one is based on the relations:

$$\begin{bmatrix} I_d \\ I_q \end{bmatrix} = \frac{1}{\sqrt{U_\alpha^2 + U_\beta^2}} \begin{bmatrix} U_\alpha & U_\beta \\ -U_\beta & U_\alpha \end{bmatrix} \begin{bmatrix} I_\alpha \\ I_\beta \end{bmatrix}. \tag{4.5}$$

The obtained signals of the flux-forming and torque-generating components of the stator current in a fixed coordinate system can be represented by the sum of constant and variable components:

$$\begin{cases} I_d = I_{d0} + \tilde{i}_d; \\ I_q = I_{q0} + \tilde{i}_q. \end{cases} \tag{4.6}$$

For the case of compensation for the variable components by current projections, the setting for the compensation in the fixed coordinate system is determined as follows:

$$\begin{bmatrix} I_{c\alpha} \\ I_{c\beta} \end{bmatrix} = \frac{1}{\sqrt{U_\alpha^2 + U_\beta^2}} \begin{bmatrix} U_\alpha & -U_\beta \\ U_\beta & U_\alpha \end{bmatrix} \begin{bmatrix} -\tilde{i}_d \\ -\tilde{i}_q \end{bmatrix}, \tag{4.7}$$

where U_α, U_β and I_α, I_β – the projections of the voltage and current vector in a fixed coordinate system; $\tilde{i}_d, \tilde{i}_q$ – the variable components of the flux-forming and torque-generating components of the stator current.

The obtained signals of compensation current by means of Clarke's direct transformations are converted into rotating coordinate system (I_{cd}, I_{cq}) and are sent to the inputs of the corresponding current regulators in the control system. The reference signals of the flux-forming and torque-generating components of the current are determined as follows:

$$\begin{cases} I'_{sd(ref)} = I_{sd(ref)} + I_{cd}; \\ I'_{sq(ref)} = I_{sq(ref)} - I_{cq}. \end{cases} \tag{4.8}$$

where $I_{sd(ref)}$ and $I_{sq(ref)}$ – the initial reference signals of the flux-forming and torque-generating components of the current, respectively; I_{cd} and I_{cq} – the compensation currents for torque-generating and flux-forming components of stator currents, respectively.

The functional diagram of the proposed vector control system with the function of compensation for the variable component of power consumption is shown in Figure 4.14.

The research of the modes of operation of the developed control system with IM series 4A112M4 with a capacity of 5.5 kW was conducted for these cases of asymmetry of the stator windings. Thus, the signals of active power consumed from the network and their spectral composition for the 4th mode are shown in Figures 4.15-4.16.

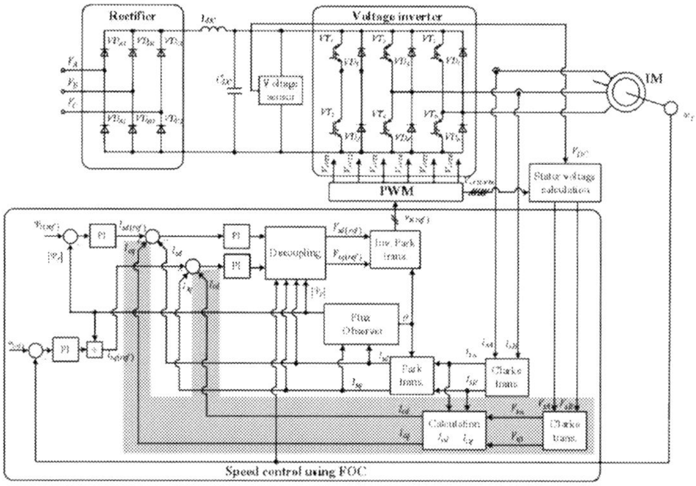

Figure 4.14. The functional diagram of the variable-frequency ED with the function of compensation for the variable component of power consumption.

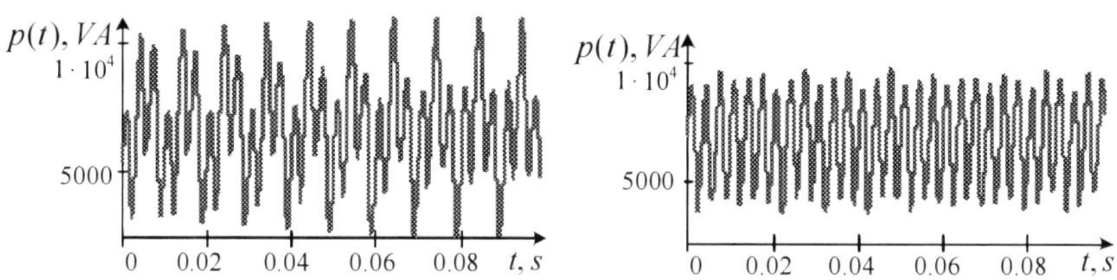

a) b)

Figure 4.15. Active power consumed from the network by asymmetric IM in the classical (a) and in the proposed (b) control system.

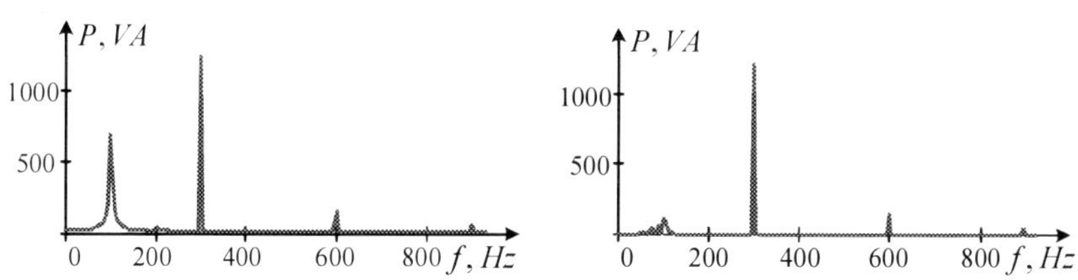

a) b)

Figure 4.16. The spectral composition of active power consumed from the network by asymmetric IM in the classical (a) and in the proposed (b) control system.

Thus, there is a significant reduction in the variable component of the active power consumed from the network. Power supply current signals are shown in Figure 4.17.

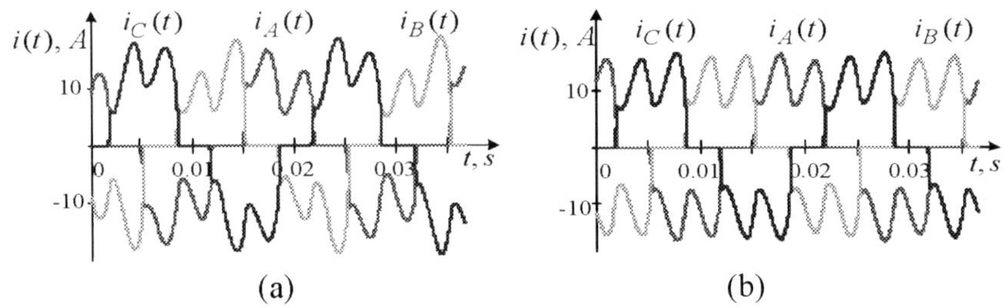

Figure 4.17. Power supply current signals during operation of asymmetric IM in the classical (a) and in the proposed (b) control system.

The signals of the active power consumed from FC and their spectral composition for the 4th mode of operation are given in Figures 4.18-4.19.

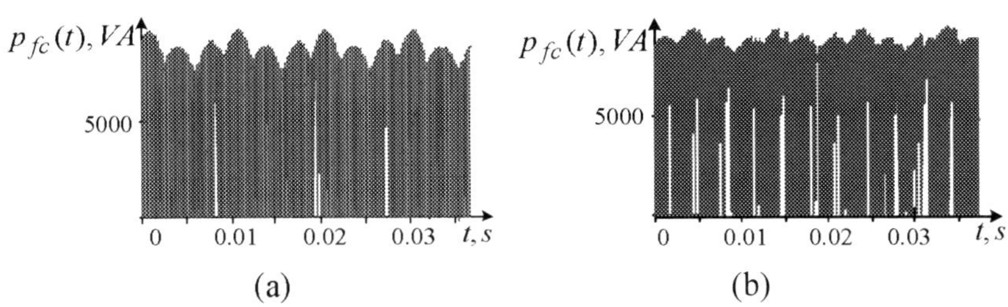

Figure 4.18. The active power consumed from FC by asymmetric IM in the classical (a) and in the proposed (b) control system.

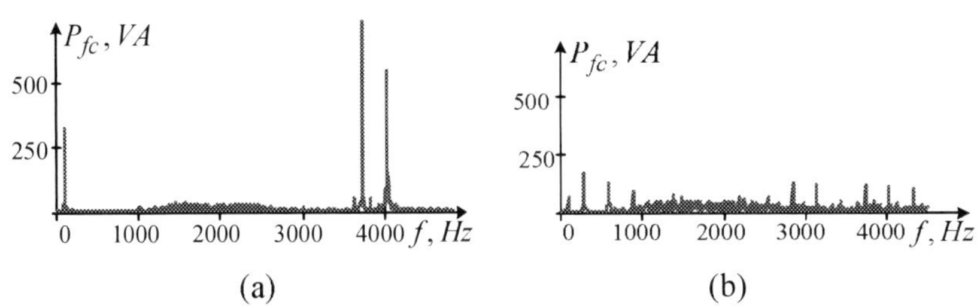

Figure 4.19. The spectral composition of the active power consumed from FC by asymmetric IM in the classical (a) and in the proposed (b) control system.

The research of the modes of operation of the proposed control system has shown that the variable component of the active power consumed from FC is also significantly reduced. The stator current signals of the researched motor are given in Figure 4.20.

A comparison of the root mean square values of the variable components of the power consumed from the network and from IM for these cases of asymmetry of the motor stator windings is given in Figure 4.21.

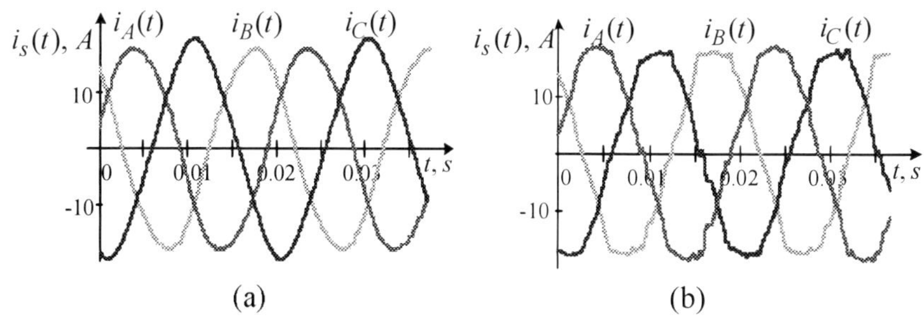

Figure 4.20. The stator current of asymmetric IM in classical (a) and in the proposed (b) control system.

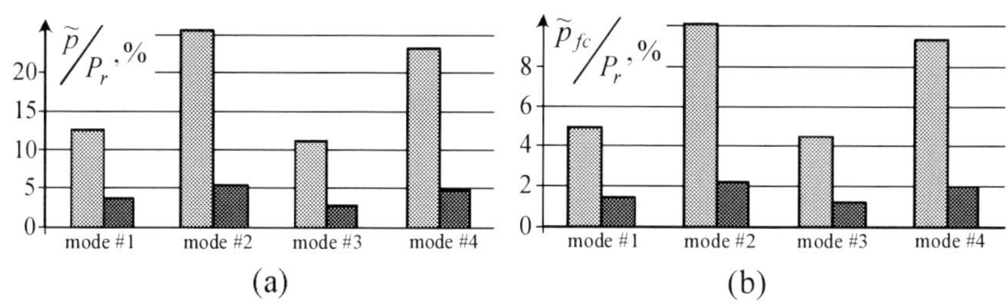

Figure 4.21. The variable components of active power consumed from the network (a) and from FC (b).

The results of the research showed that the variable component of both the power consumed from the network and the power consumed from FC is reduced by an average of 75% for cases of asymmetry in one and two phases of IM. The variable component of the electromagnetic torque in this case is not compensated for, but slightly increased (Figure 4.22).

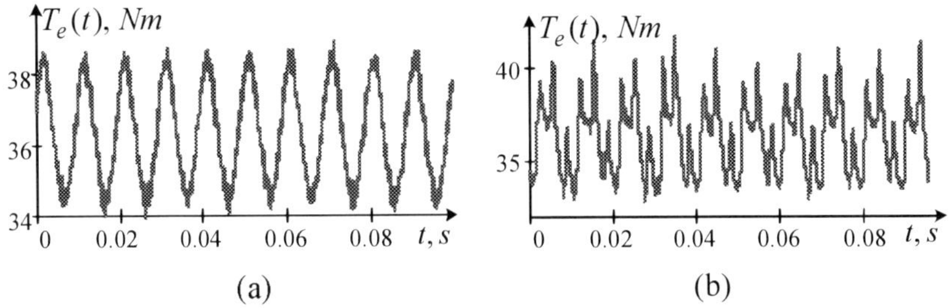

Figure 4.22. The electromagnetic torque of asymmetric IM in the classical (a) and in the proposed (b) control system.

The performed research revealed that when using the proposed control system, the variable component of the electromagnetic torque increases by 26% in asymmetry in one phase, and 24% in asymmetry in two phases of IM (Figure 4.23).

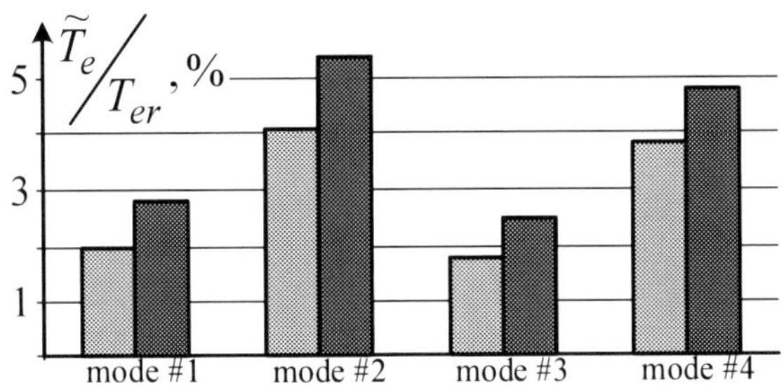

Figure 4.23. The variable components of IM electromagnetic torque.

The redistribution of losses in the stator copper by IM phases and in the semiconductor switches of FC before and after the compensation is shown in Figures 4.24-4.25.

Thus, the losses in the stator copper of the most loaded phase are reduced by an average of 51% for cases of asymmetry in both one and two phases of IM. In turn, the losses in the power diodes of the bridge rectifier of the most loaded phase are reduced by an average of 89% in asymmetry in one phase, and 72% in asymmetry in two phases; the losses in transistor switches are reduced by an average of 45% in asymmetry in one phase, and 39% in asymmetry in two phases. However, the losses in the reverse diodes of power transistors for all cases of asymmetry increase slightly, so with asymmetry in one phase they increase by 20% and in asymmetry in two phases – by 28%.

1st mode (asymmetry in phase A is 5%)

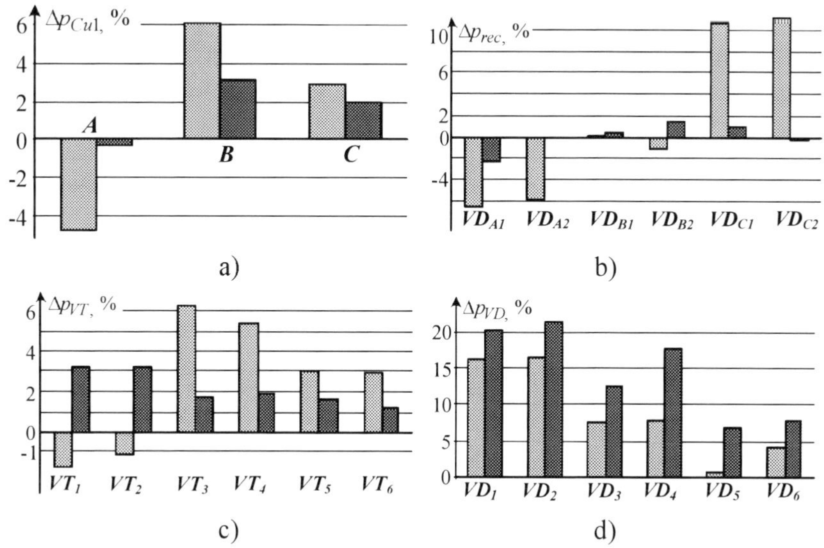

2nd Mode (Asymmetry in Phase A Is 10%)

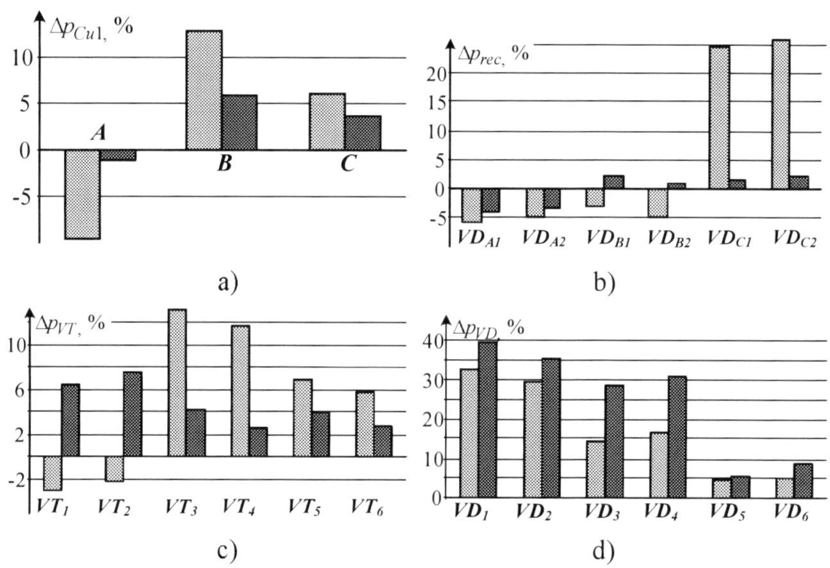

Figure 4.24. The deviation of losses in stator windings (a), input rectifier diodes (b), autonomous inverter transistors (c) and inverse diodes of transistors (d) from their rated values before and after the compensation.

3rd Mode (Asymmetry in Phase A Is 5%, in Phase C – 3%)

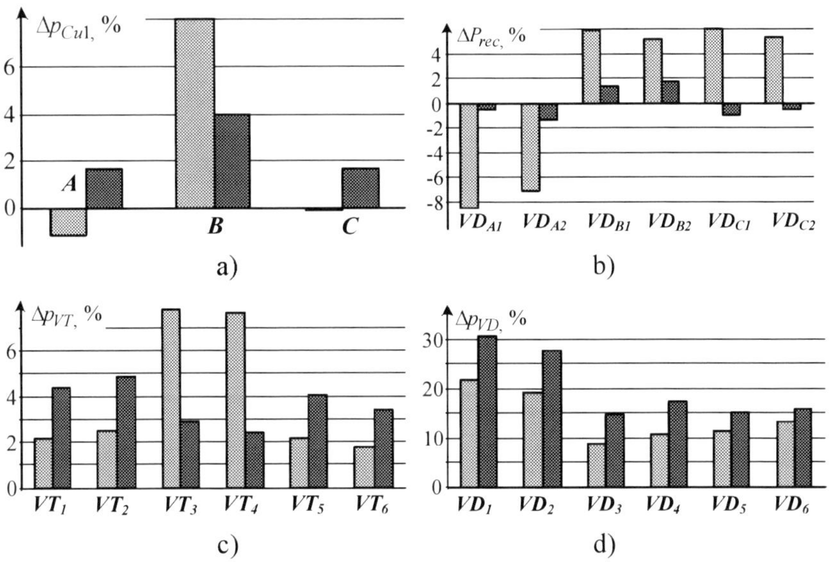

4th Mode (Asymmetry in Phase A Is 10%, in Phase C – 7%)

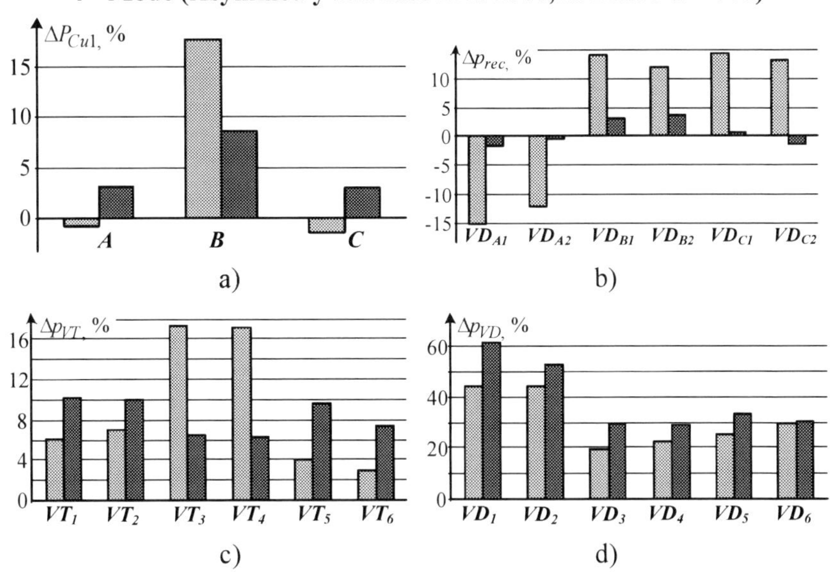

Figure 4.25. The deviation of losses in stator windings (a), input rectifier diodes (b), autonomous inverter transistors (c) and inverse diodes of transistors (d) from their rated values before and after the compensation.

The results of the research show that the use of the proposed vector control system with the function of compensation for the variable component of the three-phase active power makes it possible to significantly reduce the thermal overload of individual phases of IM and power semiconductor switches of the frequency converter. However, there is a slight increase in the variable component of the electromagnetic torque of IM.

4.3. Phase Vector Control System for an Induction Motor with Asymmetric Stator Windings

4.3.1. The Theoretical Bases of the use of the Phase Control Systems for Asymmetric Motors

In vector control systems, the electromagnetic torque of IM is determined by the product of current and flux linkage vectors. To substantiate the possibility of correcting the modes of operation of IM with asymmetric stator windings, consider the process of formation of the electromagnetic torque of IM in an analytical form in the frequency domain. The expression to determine the electromagnetic torque:

$$T_e = \frac{p_n}{\sqrt{3}}\left[(\Psi_C - \Psi_B)i_A + (\Psi_A - \Psi_C)i_B + (\Psi_B - \Psi_A)i_C\right]. \qquad (4.9)$$

where i_A, i_B, i_C – the currents of IM stator phases; Ψ_A, Ψ_B, Ψ_C – the flux linkages of the motor phases.

The following assumptions were made when considering the system of three-phase asymmetric IM windings: phase windings are powered by autonomous voltage sources u_A, u_B, u_C; the currents have only amplitude asymmetry, no phase asymmetry.

If the currents of IM phases are represented by vectors, their projections on the axis of the orthogonal coordinate system are determined as follows:

$$i_{A\alpha,\,B\alpha,\,C\alpha} = I_{A\max,\,B\max,\,C\max}\cos\left(\varphi_{iA,\,iB,\,iC}\right)$$
$$i_{A\beta,\,B\beta,\,C\beta} = I_{A\max,\,B\max,\,C\max}\sin\left(\varphi_{iA,\,iB,\,iC}\right)$$

where $I_{A\max,\,B\max,\,C\max}$ – the maximum values of the amplitudes of the stator phase currents; $\varphi_{iA,\,iB,\,iC}$ – the angles of vectors of the phases currents ($\varphi_{iA} = 0^\circ$, $\varphi_{iB} = 120^\circ$, $\varphi_{iC} = 240^\circ$).

The flux-linkage vectors projections are determined similarly:

$$\Psi_{A\alpha,\,B\alpha,\,C\alpha} = \Psi_{A\max,\,B\max,\,C\max}\cos(\varphi_{\psi A,\,\psi B,\,\psi C});$$
$$\Psi_{A\beta,\,B\beta,\,C\beta} = \Psi_{A\max,\,B\max,\,C\max}\sin(\varphi_{\psi A,\,\psi B,\,\psi C}),$$

where $\Psi_{A\max,\,B\max,\,C\max}$ – the maximum values of the amplitudes of phase flux linkages; $\varphi_{\psi A,\,\psi B,\,\psi C}$ – the angles between the vectors of flux linkages and the vectors of torque-generating components of currents.

On the basis of an automated algorithm for discrete convolution of two series [204], in which the orthogonal components of currents and flux linkages in the frequency domain are set by the first harmonics, the frequency orthogonal components of the IM electromagnetic torque are determined:

$$\begin{cases} T_{ea\alpha} = \left(\sum_{k=0}^{K-1} if\left(k-m\geq 0, i_{A\alpha_m}\Psi_{CB\alpha_{k-m}}, i_{A\alpha_m}\Psi_{CB\alpha_{m-k}}\right)\right) + \\ + \left(\sum_{k=0}^{K-1} if\left(k-m\geq 0, i_{A\beta_m}\Psi_{CB\beta_{k-m}}, -i_{A\beta_m}\Psi_{CB\beta_{m-k}}\right)\right); \\ T_{ea\beta} = \left(\sum_{k=0}^{K-1} if\left(k-m\geq 0, i_{A\beta_m}\Psi_{CB\alpha_{k-m}}, i_{A\beta_m}\Psi_{CB\alpha_{m-k}}\right)\right) + \\ + \left(\sum_{k=0}^{K-1} if\left(k-m\geq 0, i_{A\alpha_m}\Psi_{CB\beta_{k-m}}, -i_{A\alpha_m}\Psi_{CB\beta_{m-k}}\right)\right), \end{cases}$$

where K – the quantity of the harmonics of flux linkages and currents; m, k – the numbers of the harmonics of flux linkages and currents, respectively.

By summing the obtained harmonic components of the electromagnetic torque in phases, the orthogonal components of the motor torque are calculated:

$$T_{e\alpha} = T_{ea\alpha} + T_{eb\alpha} + T_{ec\alpha}; \quad T_{e\beta} = T_{ea\beta} + T_{eb\beta} + T_{ec\beta},$$

where $T_{e\alpha}$, $T_{e\beta}$ – the orthogonal components of the total electromagnetic torque of IM; $T_{ea\alpha}$, $T_{eb\alpha}$, $T_{ec\alpha}$, $T_{ea\beta}$, $T_{eb\beta}$, $T_{ec\beta}$ – the orthogonal components of the electromagnetic torque in three phases of IM.

Using the presented algorithm, the frequency orthogonal components of the electromagnetic torque of IM were obtained for the case of complete symmetry of the current and flux linkage:

$$\begin{cases} T_{e\alpha 0} = \frac{3}{2} I_{\max} \sqrt{3} \Psi_{\max}; \\ T_{e\alpha 2} = 0;\ T_{e\beta 2} = 0, \end{cases} \tag{4.10}$$

where I_{max} – the maximum value of the amplitude of the stator phase currents (for a symmetrical IM $I_{max} = I_{A\,max} = I_{B\,max} = I_{C\,max}$); Ψ_{max} – the maximum value of the amplitude of phase flux linkages (for a symmetrical IM $\Psi_{max} = \Psi_{A\,max} = \Psi_{B\,max} = \Psi_{C\,max}$); $T_{e\alpha\alpha}$ – the constant component of IM electromagnetic torque; $T_{e\alpha 2}$, $T_{e\beta 2}$ – the orthogonal components of the second harmonic of IM electromagnetic torque.

The research demonstrates that the harmonic composition of the electromagnetic torque of an IM with symmetrical stator windings contains only a constant component. To research the asymmetrical modes of operation, it is necessary to set IM asymmetrical currents, which is carried out using the asymmetry coefficient ε_i, which is determined by the ratio of the number of turns in the damaged phase to the number of turns in the undamaged phase. In accordance with this, the peak value of the asymmetrical phase current is determined as $I_{C\,max} = \varepsilon_i\, I_{max}$.

The harmonic composition of the electromagnetic torque in the presence of asymmetry of phase currents and symmetry of the IM flux linkages is determined as follows:

$$\begin{cases} T_{e\alpha 0} = I_{\max} \sqrt{3} \Psi_{\max} + \frac{1}{2} \varepsilon_i I_{\max} \sqrt{3} \Psi_{\max}; \\ T_{e\alpha 2} = \frac{1}{4} I_{\max} \sqrt{3} \Psi_{\max} - \frac{1}{4} \varepsilon_i I_{\max} \sqrt{3} \Psi_{\max}; \\ T_{e\beta 2} = -\frac{3}{4} I_{\max} \Psi_{\max} + \frac{3}{4} \varepsilon_i I_{\max} \Psi_{\max}. \end{cases} \tag{4.11}$$

Thus, with the asymmetry of IM phase currents, a variable component of the electromagnetic torque appears, as evidenced by the presence of the 2nd harmonic in the composition of the torque.

To describe the asymmetry of the flux linkages of the motor, the asymmetry coefficient ε_ψ is used, which is equal to the ratio of the amplitude value of the flux linkage in the damaged phase to the amplitude value of the flux linkage in the undamaged phase. Then the amplitude value of the flux linkage of the asymmetric phase will be determined as $\Psi_{C\max} = \varepsilon_\psi \Psi_{\max}$.

The harmonic composition of the electromagnetic torque in the presence of asymmetrical phase currents and asymmetrical motor flux linkages is determined as:

$$\begin{cases} T_{e\alpha 0} = \frac{1}{2}\varepsilon_\Psi I_{\max}\sqrt{3}\Psi_{\max} + \frac{1}{2}I_{\max}\sqrt{3}\Psi_{\max} + \\ + \frac{1}{2}\varepsilon_i I_{\max}\sqrt{3}\Psi_{\max}; \\ T_{e\alpha 2} = \frac{1}{4}I_{\max}\varepsilon_\Psi\sqrt{3}\Psi_{\max} - \frac{1}{4}\varepsilon_i I_{\max}\sqrt{3}\Psi_{\max}; \\ T_{e\beta 2} = -\frac{3}{4}I_{\max}\varepsilon_\Psi\Psi_{\max} + \frac{3}{4}\varepsilon_i I_{\max}\Psi_{\max}. \end{cases} \tag{4.12}$$

The obtained results show that under the condition of the equality of the current and flux linkage asymmetry coefficients ($\varepsilon_i = \varepsilon_\Psi$), the second harmonic of the IM electromagnetic torque is equal to zero. Thus, with a decrease in flux linkage in the asymmetrical phase of the motor, the variable component of the electromagnetic torque is compensated for.

The above analytical calculations show the possibility of compensating for the influence of damage in IM stator circuit on the dynamic and energy characteristics of ED with vector control due to the correction of the flux linkage of the asymmetric phase. However, the existing vector control systems based on IM orthogonal models do not allow carrying out actions separately on each phase. Accordingly, to eliminate this disadvantage, the control system must be implemented in a three-phase coordinate system.

4.3.2. The Features of Creating the Systems of IM Phase-by-Phase Control

As classical vector control systems in orthogonal coordinate systems do not allow influencing each phase of IM separately, a mathematical model of the phase vector control system was developed [124] to solve the problem of correcting the modes of operation of motors with asymmetric stator windings. Its functional diagram is shown in Figure 4.26.

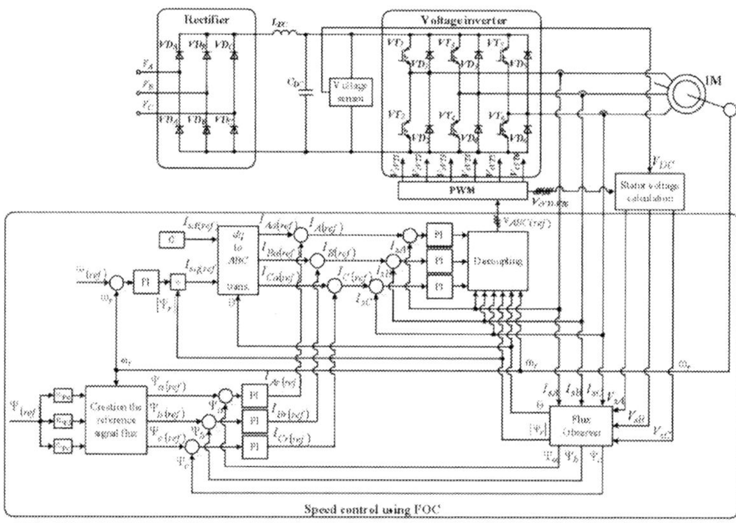

Figure 4.26. The functional diagram of tIM phase vector control system.

In the given system the circuit of the motor speed regulation is implemented in rotating coordinate system d, q. The separation of the output signal of the speed controller in the rotor flux linkage module allows obtaining the signal of the active component of the stator current (I_q^*). A transition to a fixed coordinate system is performed by means of coordinate transformations by expressions:

$$\begin{bmatrix} 0 \\ I_{qa} \\ I_{0a} \end{bmatrix} = \begin{bmatrix} \cos(\varphi_k) & \sin(\varphi_k) & 0 \\ -\sin(\varphi_k) & \cos(\varphi_k) & 0 \\ 0 & 0 & 1 \end{bmatrix} \cdot \begin{bmatrix} I_{\alpha a} \\ I_{\beta a} \\ I_{\gamma a} \end{bmatrix}. \quad (4.13)$$

The obtained reference signals of the stator current ($i^*_{\alpha a}$, $I^*_{\beta a}$) in a fixed coordinate system are converted into the reference signals of the active component of the current in a three-phase coordinate system ($I^*_{Aa}, I^*_{Ba}, I^*_{Ca}$). These transformations are performed on the basis of equations:

$$\begin{bmatrix} I_{Aa} \\ I_{Ba} \\ I_{Ca} \end{bmatrix} = \begin{bmatrix} 1 & 0 & 1 \\ -\frac{1}{2} & \frac{\sqrt{3}}{2} & 1 \\ -\frac{1}{2} & -\frac{\sqrt{3}}{2} & 1 \end{bmatrix} \cdot \begin{bmatrix} I_{\alpha a} \\ I_{\beta a} \\ I_{\gamma a} \end{bmatrix} \tag{4.14}$$

The flux linkage control circuit is performed in a fixed coordinate system, and adjustment takes place for each phase separately. In this case, the reference signal of the flux linkage must be represented by a three-phase system of sinusoidal effects with a phase shift of 120° relative to each other. For this task the block of formation of a three-phase reference signal is used (Figure 4.26). In this block, the input value is the signal of the rotor speed of the motor, which is converted into sine and cosine signals ($\sin(\gamma_r)$, $\cos(\gamma_r)$, where $\gamma_r(t) = \int \omega_r(t)dt$) with amplitudes equal to one and variable frequency [205], further conversion of which makes it possible to obtain the necessary signals represented by a three-phase system of sinusoidal influences. In turn, multiplying the flux linkage reference signal (Ψ_{ref}) by the correction factors ($\varepsilon_{\Psi a}$, $\varepsilon_{\Psi b}$, $\varepsilon_{\Psi c}$) allows obtaining a given amplitude value of flux linkage in each phase of the motor, multiplying which by the signals of the formed system of three-phase sinusoidal effects enables obtaining flux linkage reference signals ($\Psi^*_a, \Psi^*_b, \Psi^*_c$). The output signals of the flux linkage regulators are the reference signals of the reactive components of the stator phase current ($I^*_{Ar}, I^*_{Br}, I^*_{Cr}$).

Summarizing the specified active and reactive components of the stator current and using the corresponding phases of the motor we obtain the specified current signals (I^*_A, I^*_B, I^*_C), comparable to real signals (I_A, I_B, I_C). The output signals of the current regulators represent the voltage reference signals (U^*_A, U^*_B, U^*_C) of IM phases.

The presented phase control system requires information about the flux linkage of the rotor for all phases of the motor. The existing methods of indirect flux determination allow calculations only in orthogonal coordinate systems [206]. Accordingly, it was proposed to calculate the flux linkage of the rotor in a three-phase coordinate system. With this purpose, it is possible to use the equations of electric balance of the stator of an equivalent two-phase induction machine in an orthogonal coordinate system rotating at arbitrary angular velocity ω_k relative to the "natural" three-phase coordinate system:

$$\dot{U}_s = R_s\dot{I}_s + \sigma L_s \frac{d\dot{I}_s}{dt} + k_r \frac{d\dot{\Psi}_r}{dt} + j\sigma L_s \omega_k \dot{I}_s + jk_r \omega_k \dot{\Psi}_r . \qquad (4.15)$$

The solution to this equation with respect to the derivative of the flux linkage of the rotor enables obtaining the following expression:

$$\frac{d\dot{\Psi}_r}{dt} = \frac{L_r}{L_\mu}\left(\dot{U}_s - R_s\dot{I}_s - \sigma L_s \frac{d\dot{I}_s}{dt} - j\sigma L_s \omega_k \dot{I}_s\right) - j\omega_k \dot{\Psi}_r . \qquad (4.16)$$

Taking into account that the observer is performed in a fixed coordinate system, i.e., $\omega_k = 0$, the given equation takes the form:

$$\frac{d\dot{\Psi}_r}{dt} = \frac{L_r}{L_\mu}\left(\dot{U}_s - R_s\dot{I}_s - \sigma L_s \frac{d\dot{I}_s}{dt}\right). \qquad (4.17)$$

As the generalized vectors can be decomposed into components by individual phases [126], the equation of the rotor flux linkage model looks as follows:

$$\begin{cases} \dfrac{d\Psi_a}{dt} = \dfrac{L_a}{L_\mu}\left(U_A - R_A I_A - \sigma L_A \dfrac{dI_A}{dt}\right); \\ \dfrac{d\Psi_b}{dt} = \dfrac{L_b}{L_\mu}\left(U_B - R_B I_B - \sigma L_B \dfrac{dI_B}{dt}\right); \\ \dfrac{d\Psi_c}{dt} = \dfrac{L_c}{L_\mu}\left(U_C - R_C I_C - \sigma L_C \dfrac{dI_C}{dt}\right). \end{cases} \qquad (4.18)$$

Taking into account the fact that when creating a vector control system it is necessary to have information not only about the components of the flux linkage vectors, but also about its module, in this case the flux linkage module of IM rotor can be calculated by expression [161]:

$$|\Psi_r| = \sqrt{\frac{2}{3}\left(\Psi_a^2 + \Psi_b^2 + \Psi_c^2\right)}. \tag{4.19}$$

Construction of a control system in a three-phase coordinate system imposes certain restrictions on the initial characteristics of ED systems with vector control. Accordingly, in order to be able to provide high dynamic characteristics of the ED system with phase control, and independent control of the flux linkage of the rotor and the electromagnetic torque of the motor, it is necessary to provide compensation for cross-connections according to the stator current projections [207]. With this purpose in view we consider the principles of creating a system of compensation for cross-links in classical vector control systems.

Thus, in vector control system (VCS) the independent control of the flux linkage of the rotor and the electromagnetic torque (speed) of the motor is achieved by compensating for the mutual influence of the projections of the stator current vector I_d and I_q in the rotating coordinate system, which is characterized by the presence of cross-links of these projections in IM mathematical model. This problem is solved with the help of the compensation unit, which performs the separation of the control channels by neutralizing the effects of internal feedback. This compensation is performed by introducing the same cross-links to the inverter voltage input as in the mathematical model of the motor, but taken with opposite signs [140]:

$$U_{kd} = \left(-1/k_{fc}\right)I_d k_1;\; U_{kq} = \left(1/k_{fc}\right)\left(I_q k_1 + k_2\right), \tag{4.20}$$

where U_{kd}, U_{kq} – compensation voltage; k_{fc} – FC transfer coefficient; I_d, I_q – the projections of the stator current vector on the axis of the rotating coordinate system *d, q*; $k_1 = \sigma L_s p_n \omega_r$; $k_2 = p_n \omega_r \Psi_r \left(L_\mu / L_r\right)$.

Using expressions of conversion from a rotating coordinate system to a fixed one, the mathematical description of the compensation unit can be written as follows:

$$\begin{cases} U_{k\alpha} = \dfrac{1}{k_{fc}}\left(-I_d k_1 \cos(\theta) - \left(I_q k_1 + k_2\right)\sin(\theta)\right); \\ U_{k\beta} = \dfrac{1}{k_{fc}}\left(-I_d k_1 \sin(\theta) + \left(I_q k_1 + k_2\right)\cos(\theta)\right), \end{cases} \tag{4.21}$$

where θ – the angle of rotation of the fixed coordinate system relative to the rotating one, which is determined in the control system.

Further conversion of compensation signals from a fixed coordinate system to a three-phase one allows obtaining the following expressions:

$$\begin{cases} U_{kA} = \dfrac{1}{k_{fc}}\left(-I_d k_1 \cos(\theta) - \left(I_q k_1 + k_2\right)\sin(\theta)\right); \\ U_{kB} = \dfrac{1}{k_{fc}}\left(\dfrac{1}{2} I_d k_1 \cos(\theta) + \dfrac{1}{2}\left(I_q k_1 + k_2\right)\sin(\theta)\right) + \\ + \dfrac{\sqrt{3}}{2}\left(-I_d k_1 \sin(\theta) + \left(I_q k_1 + k_2\right)\cos(\theta)\right); \\ U_{kC} = \dfrac{1}{k_{fc}}\left(\dfrac{1}{2} I_d k_1 \cos(\theta) + \dfrac{1}{2}\left(I_q k_1 + k_2\right)\sin(\theta)\right) - \\ - \dfrac{\sqrt{3}}{2}\left(-I_d k_1 \sin(\theta) + \left(I_q k_1 + k_2\right)\cos(\theta)\right). \end{cases} \tag{4.22}$$

In turn, the projections of the stator current on the axes of the rotating coordinate system can be obtained on the basis of a direct transition from a three-phase coordinate system to orthogonal one:

$$\begin{cases} I_d = \dfrac{2}{3}\left(\begin{array}{l} I_A \cos(\theta) + I_b\left(-\dfrac{1}{2}\cos(\theta) + \dfrac{\sqrt{3}}{2}\sin(\theta)\right) + \\ + I_c\left(-\dfrac{1}{2}\cos(\theta) - \dfrac{\sqrt{3}}{2}\sin(\theta)\right) \end{array}\right); \\ I_q = \dfrac{2}{3}\left(\begin{array}{l} I_A \sin(\theta) + I_b\left(-\dfrac{1}{2}\sin(\theta) - \dfrac{\sqrt{3}}{2}\cos(\theta)\right) + \\ + I_c\left(-\dfrac{1}{2}\sin(\theta) + \dfrac{\sqrt{3}}{2}\cos(\theta)\right) \end{array}\right). \end{cases} \tag{4.23}$$

Substituting (4.23) in (4.22), we obtain expressions for compensation of cross-links by projections of the current of IM stator in the three-phase coordinate system. Thus, the presented mathematical description of the compensation unit can be used in phase vector control systems.

In turn, the representation of IM by a mathematical model in the rotating coordinate system, when *d*-axis is directed by the spatial vector of rotor flux linkage, and compensation for the cross-links of the stator current projections make it possible to adjust the classical vector control systems according to the principles of subordinate coordinate control [208]. Thus, in vector control, the synthesis of regulators is usually based on the consideration of a simplified linearized mathematical model of the motor. Moreover, the regulators of the torque-generating and current-forming components of the stator current, as well as the regulator of the flux linkage of the rotor, are usually synthesized to the modular optimum, and the speed controller, depending on the need to provide the system with astatic properties, is synthesized to modular or symmetric optimum.

The classical approaches to the construction of vector control systems on the principles of subordinate control involve the operation of projections of the stator current and flux linkage of the rotor in a rotating coordinate system, i.e., with DC voltage signals. However, in phase control, the regulation of these variables occurs in a fixed coordinate system, i.e., the control system operates with periodic signals. Accordingly, there is a problem of the possibility of using classical approaches to the synthesis of regulators of the vector control system in the regulation of periodic signals. With this pupose in view, the method of the synthesis of closed-loop controllers of the classical vector control system was considered.

The transfer function of the modular optimum closed-loop current control circuit is determined as follows:

$$W_c^{cl}(p)=\frac{\left(k_{pc}p+k_{ic}\right)k_{fc}\left(1/R_e\right)\left(T_c^s p+1\right)}{\left(k_{pc}p+k_{ic}\right)k_{fc}\left(1/R_e\right)k_c^s+p\left(T_{fc}p+1\right)\left(T_s p+1\right)\left(T_c^s p+1\right)}, \tag{4.24}$$

where T_{fc}, T_c^s, T_s – the time constant of the frequency converter, the current sensor and the equivalent time constant of the stator circuit; k_{pc}, k_{ic} – the coefficients of proportional and integral component of the current regulator,

k_{fc}, k_c^s – the transmission coefficients of frequency converter and current sensor;

R_e – total resistance of IM phase.

The optimized current control circuit is a first-order astatic control system. At the frequency of modulation of FC power switches equal to 8 kHz such adjustment of the system allows receiving the following indicators of the operation quality: static control error $\Delta I = 0$; over-regulation $\sigma = 5.3$ %; time of the first adjustment $t_1 = 0.36 \cdot 10^{-3}$ s; regulation time $t_2 = 0.68 \cdot 10^{-3}$ s.

The quality of transients in control systems with periodic signals can be determined on the basis of indirect criteria, which include the frequency characteristics of the system [209]. Accordingly, on the basis of the presented transfer functions the amplitude-frequency characteristic (AFC) and phase-frequency characteristics (PFC) of the current control circuit were constructed. So, Figure 4.27 contains AFC and PFC of the open and closed circuit of current regulation at the frequency converter frequency of modulation equal to 8 kHz.

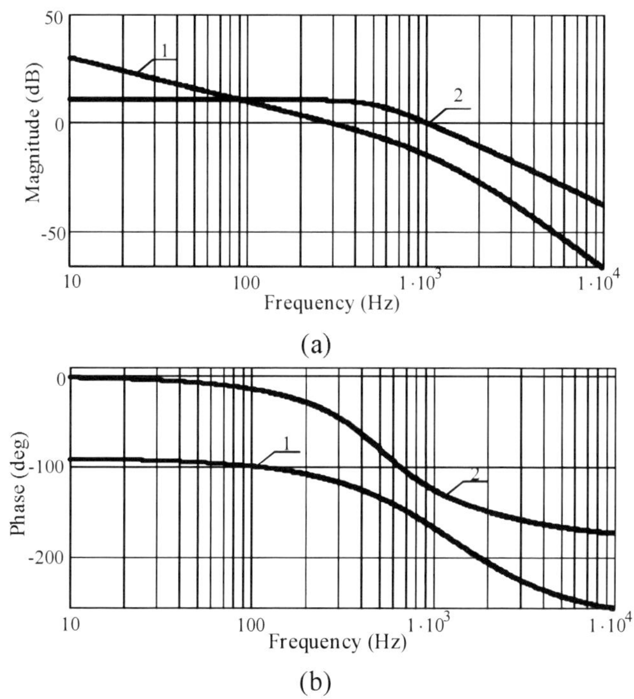

Figure 4.27. AFC(a) and PFC (b) of the open (1) and closed (2) circuit of current regulation.

The quality of control in the current control circuit is assessed on the basis of such criteria as the stability margin by the amplitude (L_{sm}) and phase (φ_{sm}), cut-off frequency ($f_{cut-off}$) – for the open circuit of current regulation; system oscillation index (M) and control system bandwidth (f_p – for the closed regulation circuit. The results of the research are given in Table 4.1.

Table 4.1. The control quality indicators in the current regulation circuit

f_{pwm}, kHz	Regulation circuit			
	Open			Closed
	$f_{cut-off}$, Hz	L_{sm}, dB	φ_{sm}, deg	f_p, Hz
2	318.437	18.069	63.107	141.807
4	636.715	18.064	62.732	283.773
8	1273.27	18.062	63.107	560.305
16	254.48	18.062	63.094	1118.54

The conducted research has shown that the indicator of oscillation of the closed circuit of current regulation for the considered frequencies of modulation of power switches of the autonomous voltage inverter is equal to one. Accordingly, the closed control circuit is not subject to the oscillatory nature of the transients.

When adjusting periodic signals, the amplitude and phase of the signal can be used as evaluation criteria. Therefore, based on the calculated AFC and PFC, the errors of signal control in the closed circuit of current control were determined (Table 4.2).

Table 4.2. Errors in the regulation of periodic signals in a closed stator current circuit

f_{pwm}, kHz	Errors, %	Frequency, Hz			
		25	50	75	100
2	A	0.2467	0.243	2.104	9.011
	φ	3.803	7.899	12.433	17.232
4	A	0.0733	0.247	0.381	0.243
	φ	1.882	3.803	5.8	7.899
8	A	0.019	0.073	0.154	0.247
	φ	0.938	1.882	2.835	3.803
16	A	0.0048	0.019	0.042	0.073
	φ	0.469	0.938	1.409	1.882

When optimizing the flux linkage control circuit, the internal current control circuit is usually represented by the transfer function of the optimized circuit. Accordingly, the transfer function of the modular optimized closed control circuit is determined as follows:

$$W_{fl}^{cl}(p) = \frac{\left(k_{pfl}p + k_{ifl}\right)W_c^{cl}L_\mu\left(T_{fl}^s p + 1\right)}{\left(k_{pfl}p + k_{ifl}\right)W_c^{cl}L_\mu k_{fl}^s + p\left(T_r p + 1\right)\left(T_{fl}^s p + 1\right)}, \tag{4.25}$$

where T_{fl}^s, T_r – time constant of the flux linkage and rotor circuit; k_{pfl} k_{ifl} – the coefficients of proportional and integral component of the flux linkage regulator; k_{fl}^s – the transfer coefficient of the flux linkage sensor.

The optimized flux linkage control circuit is a first-order astatic control system. At the frequency of modulation of the power switches of the autonomous voltage inverter of the frequency converter equal to 8 kHz, such adjustment of the system makes it possible to obtain the following indicators of operation quality: static control error $\Delta\psi = 0$; over-regulation $\sigma = 0.91$ %; the time of the first adjustment – $t_1 = 1.1 \cdot 10^{-3}$ s; regulation time – $t_2 = 1.84 \cdot 10^{-3}$ s.

The control quality indicators in the flux control circuit are given in Table 4.3. The oscillation index of the closed flux linkage control circuit for the considered frequencies of modulation of the frequency converter is equal to one. AFC and PFC of the flux linkage control circuit are shown in Figure 4.28.

Table 4.3. The control quality indicators in the flux linkage control circuit

f_{pwm}, kHz	**Regulation circuit**			
	Open			**Closed**
	$f_{cut\text{-}off}$, **Hz**	L_{sm}, **dB**	φ_{sm}, **deg**	f_p, **Hz**
2	95.643	13.15	66.18	51.998
4	191.288	13.15	66.325	104.624
8	382.561	13.15	66.253	208.779
16	764.978	13.147	66.775	418.02

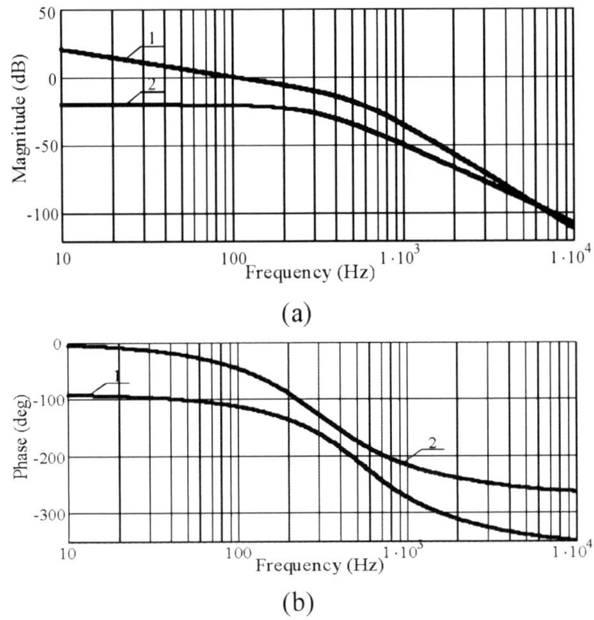

Figure 4.28. AFC (a) and PFC (b) of the open (1) and closed (2) circuit of regulation of flux linkage.

The errors in the regulation of periodic signals in a closed circuit are given in Table 4.4.

Table 4.4. The errors in the regulation of periodic signals in a closed circuit of the rotor flux linkage

f_{pwm}, kHz	Errors, %	Frequency, Hz			
		25	50	75	100
2	A	6.564	26.883	51.493	69.69
	φ	12.571	24.828	35.034	42.53
4	A	1.572	6.564	15.252	26.88
	φ	6.263	12.571	18.839	24.83
8	A	0.387	1.572	3.61	6.564
	φ	3.127	6.263	9.412	12.57
16	A	0.097	0.387	0.877	1.572
	φ	1.563	3.127	4.693	6.263

The performed research confirms the possibility of using classical approaches to the synthesis of regulators of the vector control system in the regulation of periodic signals. It is shown that the errors of regulation by the amplitude and the phase of the signals in the closed circuits of the regulation of current and flux linkage depend on the frequency of modulation of the power switches of the autonomous voltage inverter.

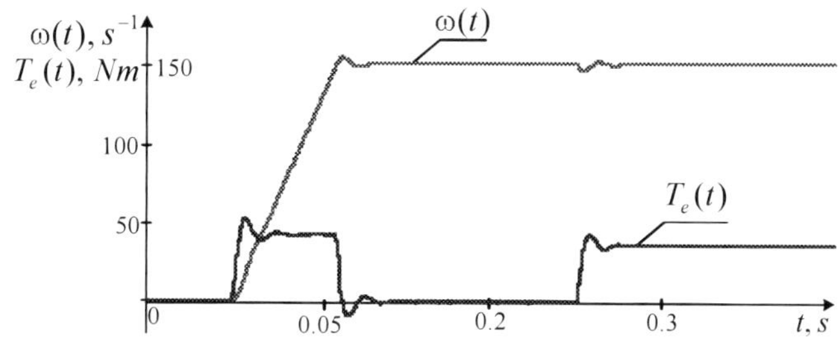

Figure 4.29. The transients in the speed and torque of IM as a part of ED with system of phase vector control.

The research of the modes of operation of the variable-frequency ED with the presented phase vector control system was carried out for the motor of the 4A112M4 series with a power of 5.5 kW. The graphs of transients according to the speed and electromagnetic torque of symmetric IM are given in Figure 4.29. To reduce the errors of periodic signal control in the closed circuits of the stator current control and rotor flux linkage during modeling, the modulation frequency of the power switches of the autonomous voltage inverter was chosen equal to 8 kHz.

The reference signals of the active and reactive components of the symmetric IM current, formed in the control system, are shown in Figure 4.30.

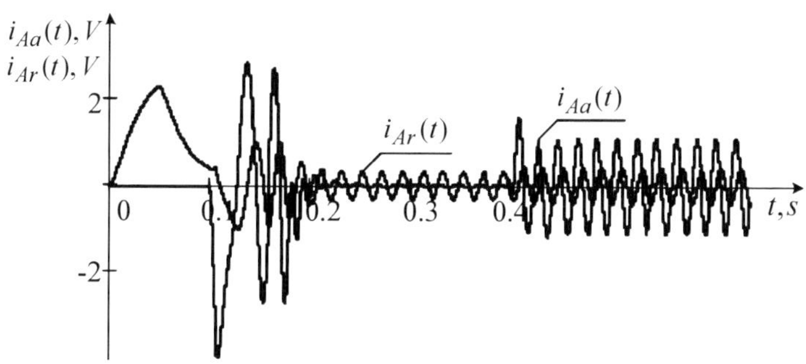

Figure 4.30. The reference signals of the active and reactive components of a symmetrical IM stator current.

The research of the modes of operation of the developed control system has shown that the dynamic and energy characteristics of phase vector control systems in a wide range of speed control coincide with the corresponding parameters of the classical control system of symmetric IM. In turn, during the operation of the proposed control system with an asymmetric motor, there is a slight violation of the symmetry of the stator voltage, the appearance of asymmetry of the stator current, and the flux linkage of the rotor in this case remains symmetrical (Figure 4.31). Hodographs of these signals are given for the case of asymmetry in the motor phase A equal 10%.

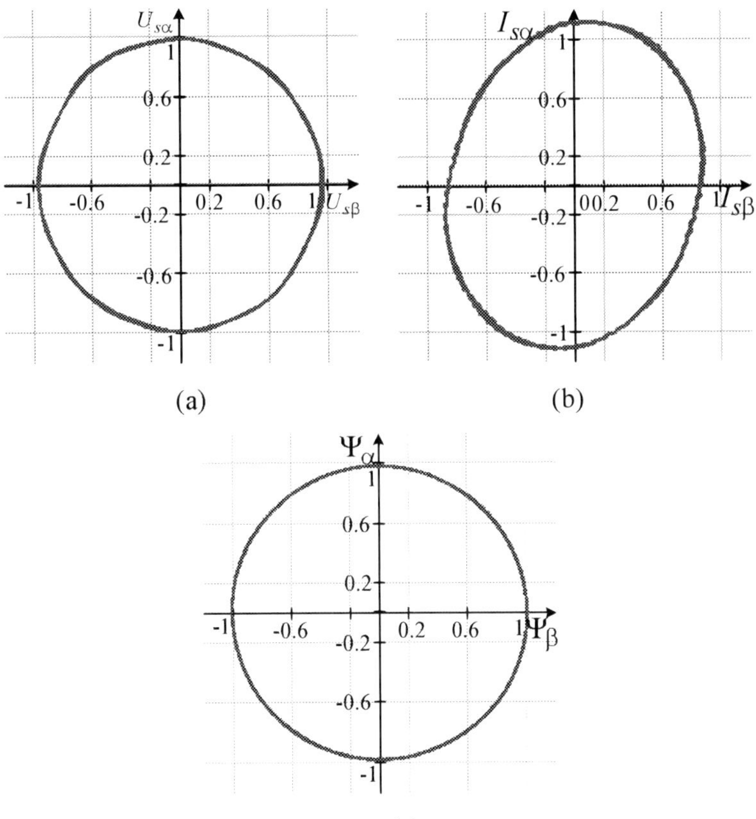

Figure 4.31. The hodographs of voltage (a), current (b) and flux linkage of the rotor (c) of asymmetric IM in the phase vector control system.

Separate control of the reactive energy transmission channel allows changing the reactive (magnetizing) components of the currents in the individual phases of IM stator. Varying the signal of the flux linkage in one phase of the control system, changes the magnetizing component of the current, and, in fact, the voltage of the corresponding phase of the motor. Thus, an asymmetric supply voltage system is formed, which accordingly affects the processes of energy conversion in IM itself.

The research of the modes of operation of the proposed control systems has shown that the effect of asymmetry of motor windings on the energy and dynamic characteristics of ED systems can be compensated for by correcting the signal of the flux linkage in the asymmetric phase of the motor [210]. In this case, the optimal value of the flux linkage of the asymmetric phase differs depending on the adjustment of the control system to compensate for the variable component of the electromagnetic torque or to compensate for the variable component of the power consumed by the motor.

4.3.3. The Adjustment of the Phase Control System to Compensate for the Variable Component of IM Electromagnetic Torque

In the mode of the compensation for the variable component of the electromagnetic torque of the asymmetric motor, the relationship between the asymmetry of the stator windings and the optimal value of the flux linkage of the rotor of the asymmetric phase can be represented as follows [123]:

$$\varepsilon_{\Psi} = \varepsilon_{w} \tag{4.26}$$

where ε_{w} – the coefficient of the stator windings asymmetry; ε_{Ψ} – the correction coefficient of the flux linkage of IM asymmetric phase.

The research of the influence of variation of the reference value of flux linkage of IM asymmetric phase on the value of the variable component of the electromagnetic torque of the motor is given in Figure 4.32. The results are presented for cases of asymmetry in phase A equal to 4%, 6%, 8% and 10%.

The research demonstrates that the reduction of the flux linkage of the asymmetric phase in proportion to the asymmetry of the active resistances in IM phases results in a decrease in the variable component of the electromagnetic torque to an acceptable level. Thus, when the flux linkage of the asymmetric phase changes, an asymmetric flux linkage of the motor is formed (Figure 4.33), due to which the variable component of the electromagnetic torque is compensated for at asymmetric currents.

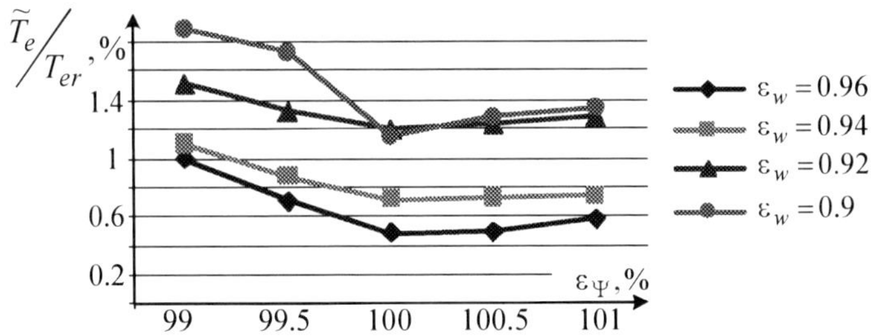

Figure 4.32. The variable components of IM electromagnetic torque at the variation of the reference value of the flux linkage.

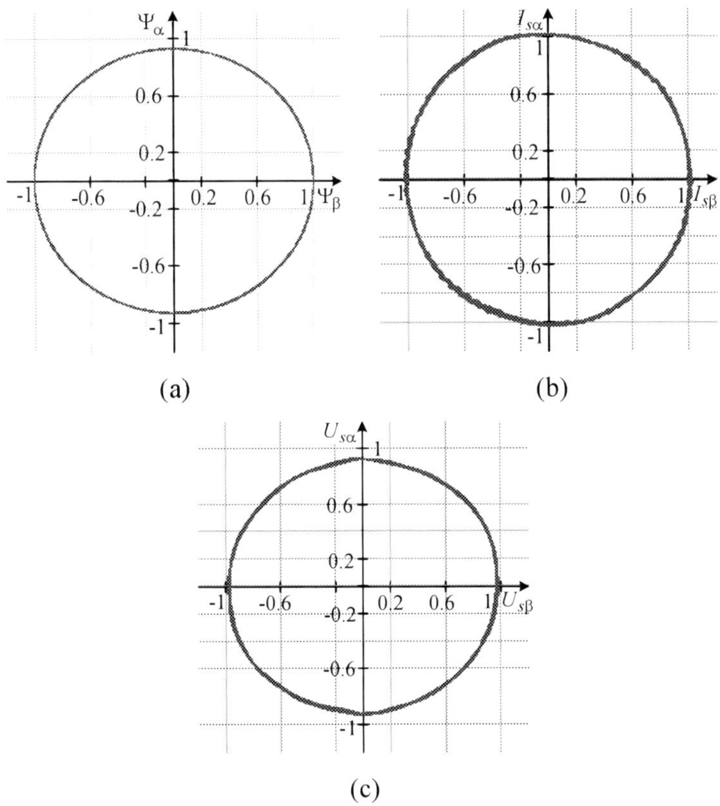

Figure 4.33. The hodographs of voltage (a), current (b) and flux linkage of IM rotor (c) with asymmetry in phase A equal to 10% with adjustment of the phase vector control system to compensate for the variable component of the electromagnetic torque.

In the considered range of changes of the reference value of flux linkage of motor asymmetric phase the variable components of the active power consumed from a network and from FC increase (Figure 4.34). The redistribution of losses in the stator copper by IM phases during the adjustment of the phase control system to compensate for the variable component of the electromagnetic torque is shown in Figure 4.35.

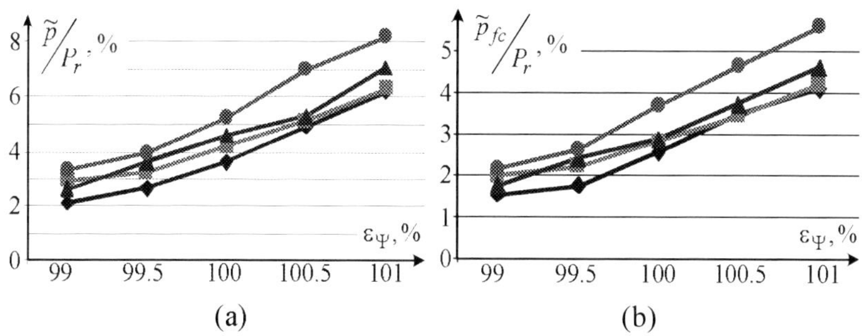

Figure 4.34. The variable components of the active power consumed from the network (a) and from FC (b) at the variation of the reference value of the flux linkage.

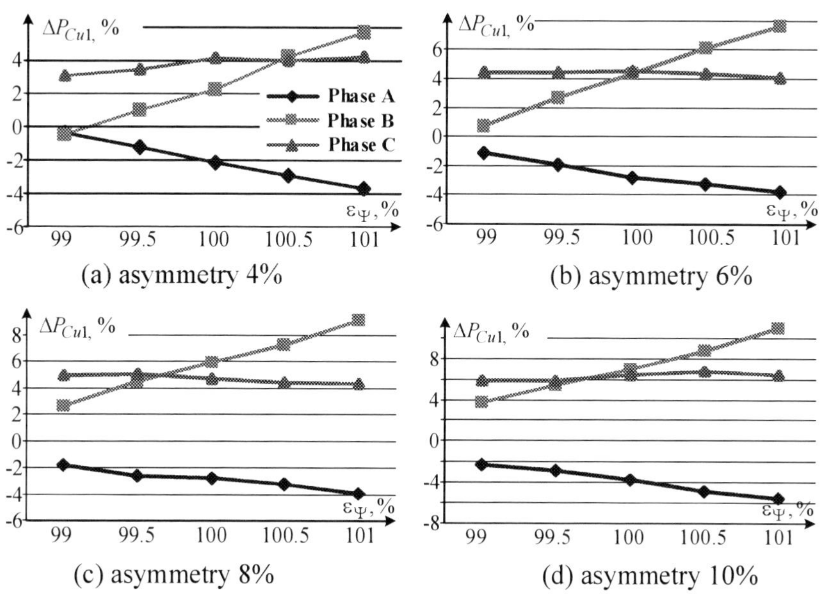

Figure 4.35. The stator copper losses by the phases of asymmetric IM at the variation of the reference value of the flux linkage.

The research of the operating modes of the phase vector control system with asymmetric IM was conducted for four operating modes. Thus, the graphs of transients according to the electromagnetic torque of the studied IM in the proposed system of phase vector control are presented in Figure 4.36, and their spectral composition is shown in Figure 4.37. The signals are presented for the 4th operating mode.

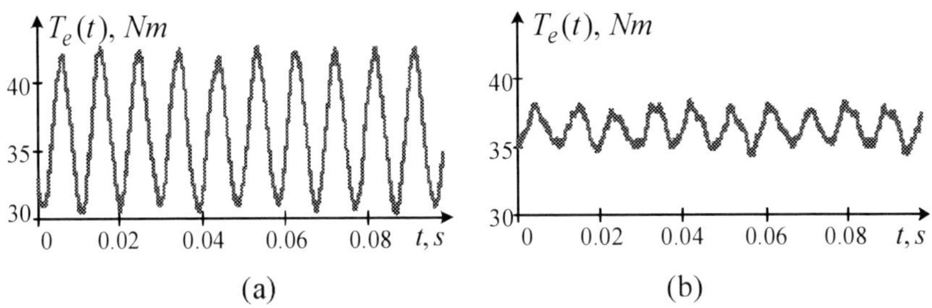

Figure 4.36. The electromagnetic torque of asymmetric IM before (a) and after (b) the compensation.

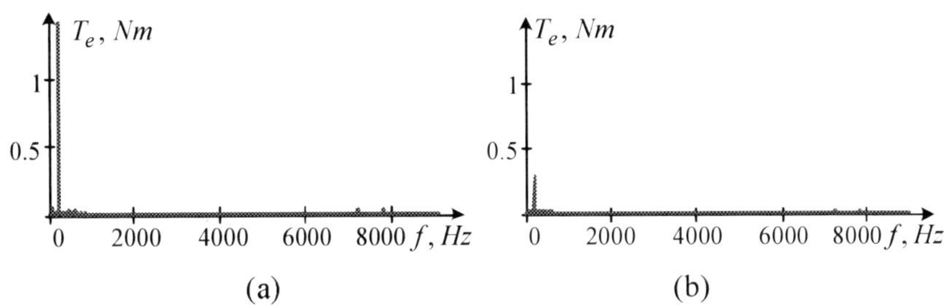

Figure 4.37. The spectral composition of IM electromagnetic torque before (a) and after (b) the compensation.

The comparison of the variable components of the electromagnetic torque before and after compensation for these cases of asymmetry of IM stator windings is given in Figure 4.38. Thus, the use of a phase vector control system with the adjustment to compensate for the variable component of the electromagnetic torque makes it possible to reduce it by an average of 83% in asymmetry of one IM phase, and 77% in asymmetry in two phases.

The redistribution of the losses in the stator copper by IM phases and in FC power elements with these cases of asymmetry of the windings before and after compensation is shown in Figures 4.39-4.40.

Thus, during the adjustment of the phase vector control system to compensate for the variable component of the electromagnetic torque, the loss in the stator copper of the most loaded phase can be reduced by an average of 40% for asymmetry in one phase and by 13% for asymmetry in two phases. In turn, the losses in the power diodes of the input rectifier of the most loaded phase are reduced by an average of 46% in asymmetry in one phase, and by 27% in asymmetry in two phases; the losses in transistor switches are reduced by an average of 62% in asymmetry in one phase, and by 36% in asymmetry in two phases; there is an increase in losses in the reverse diodes of power transistors by 31% with asymmetry in one phase and by 32% with asymmetry in two IM phases.

The results of the performed research showed that when adjusting the control system to compensate for the variable component of the electromagnetic torque of the motor, there is also a decrease in the variable component of the active power consumed from the network and from FC. Thus, the signals of active power consumed from the network and their spectral composition for the 4th mode are shown in Figures 4.41-4.42.

Figure 4.38. The variable components of IM electromagnetic torque.

1st mode (asymmetry in phase A is 5%)

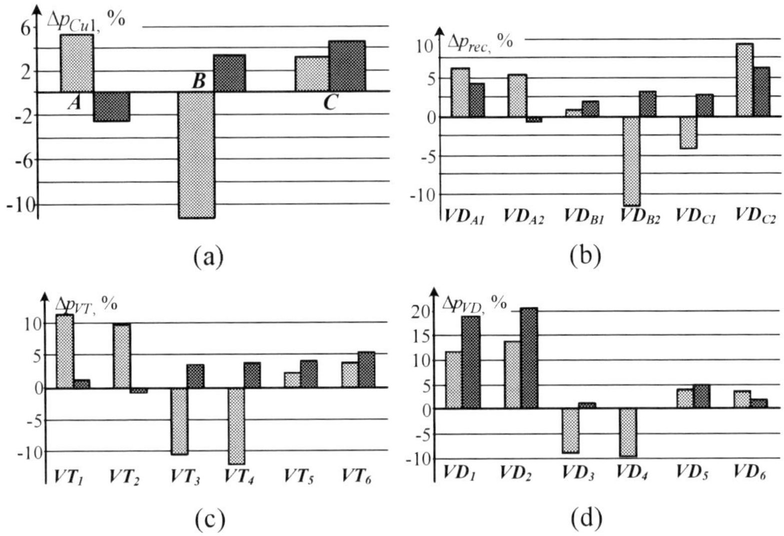

2nd mode (asymmetry in phase A is 10%)

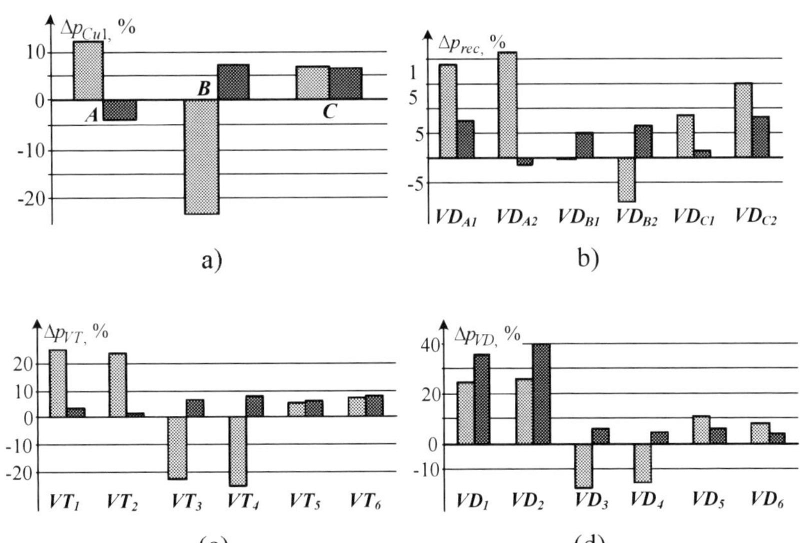

Figure 4.39. The deviation of losses in stator windings (a), input rectifier diodes (b), autonomous inverter transistors (c) and inverse diodes of transistors (d) from their rated values before and after compensation.

3rd mode (asymmetry in phase A is 5%, in phase C – 3%)

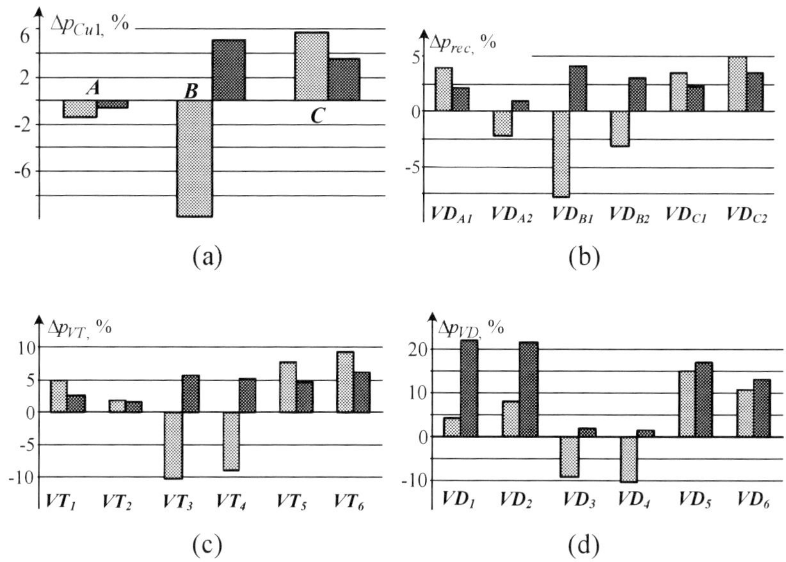

4th mode (asymmetry in phase A is 10%, in phase C – 7%)

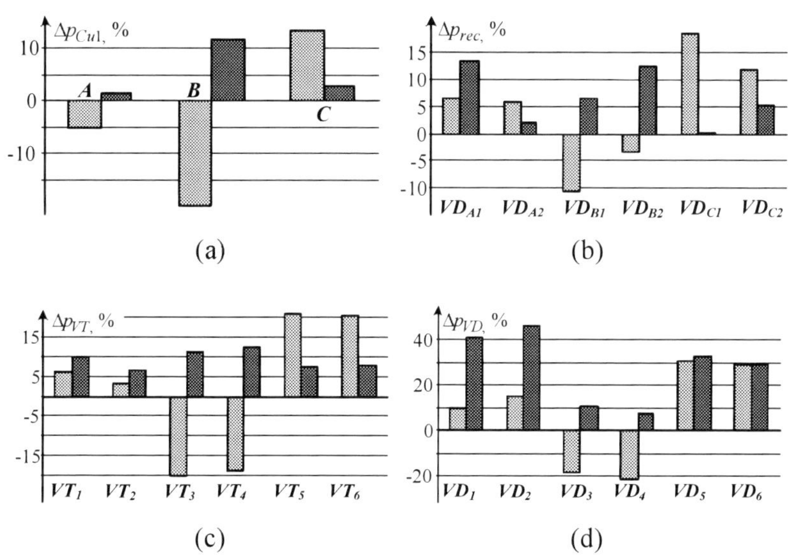

Figure 4.40. The deviation of losses in stator windings (a), input rectifier diodes (b), autonomous inverter transistors (c) and inverse diodes of transistors (d) from their rated values before and after compensation.

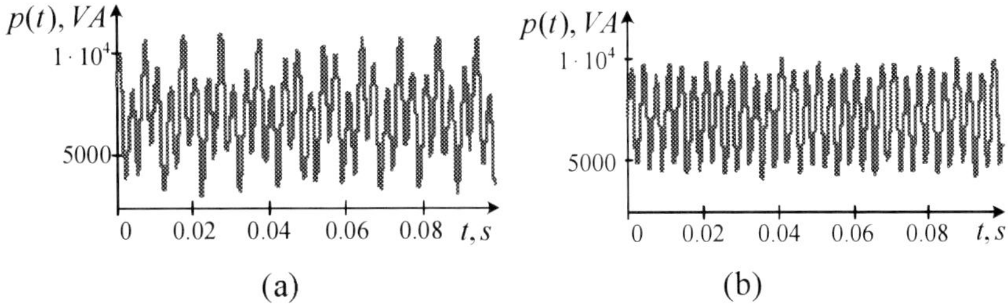

Figure 4.41. The active power consumed from the network by an asymmetric IM before (a) and after (b) compensation.

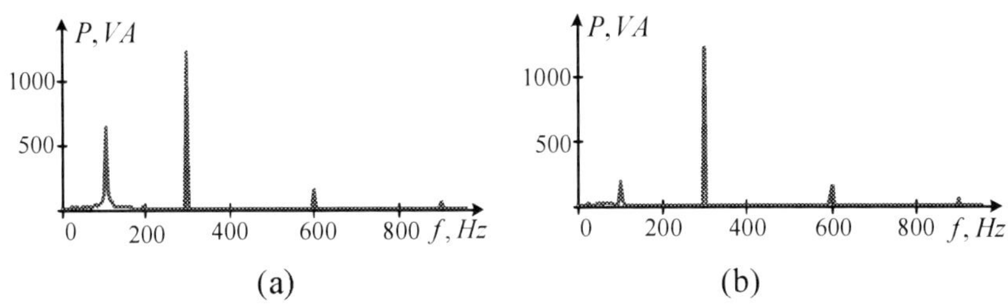

Figure 4.42. The spectral composition of the active power consumed from the network by an asymmetric IM before (a) and after (b) compensation.

The signals of current consumption from the network before and after the compensation are presented in Figure 4.43.

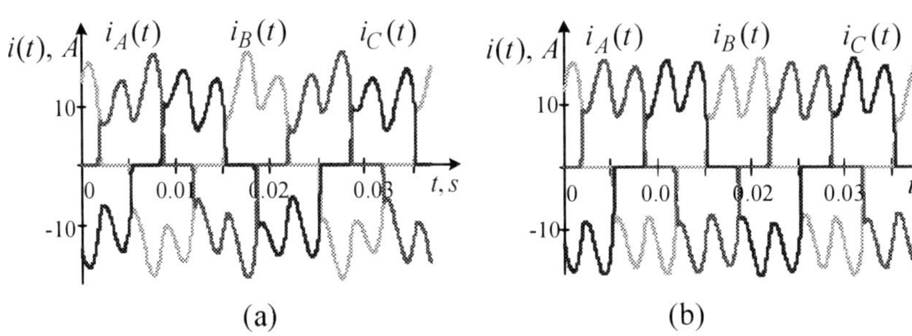

Figure 4.43. The current signals consumed from the network during the operation of an asymmetric IM before (a) and after (b) compensation.

The signals of the active power consumed from FC and their spectral composition for the 4th mode of operation are given in Figures 4.44-4.45.

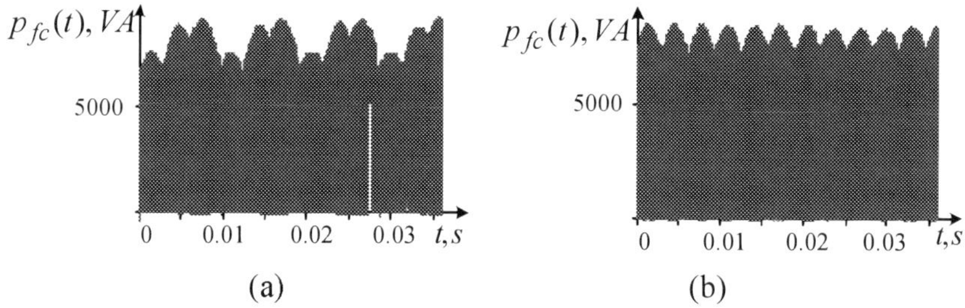

Figure 4.44. The active power consumed from FC of asymmetric IM before (a) and after (b) compensation.

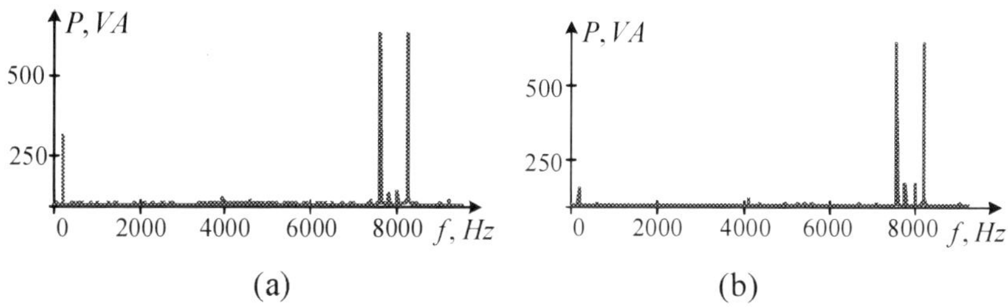

Figure 4.45. The spectral composition of active power of an asymmetric IM consumed from FC before (a) and after (b) compensation.

The stator current signals of the asymmetric IM before and after compensation for the 4th operating mode are given in Figure 4.46.

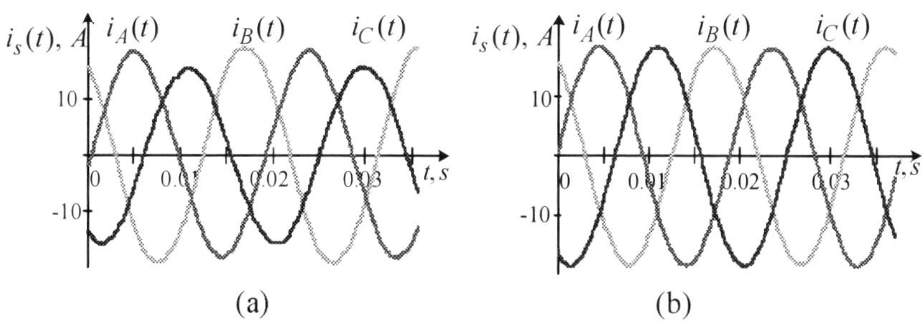

Figure 4.46. The stator current of asymmetric IM before (a) and after (b) compensation.

A comparison of the root mean square values of the variable components of the power consumed from the network and from FC for these cases of asymmetry of the motor stator windings is given in Figure 4.47.

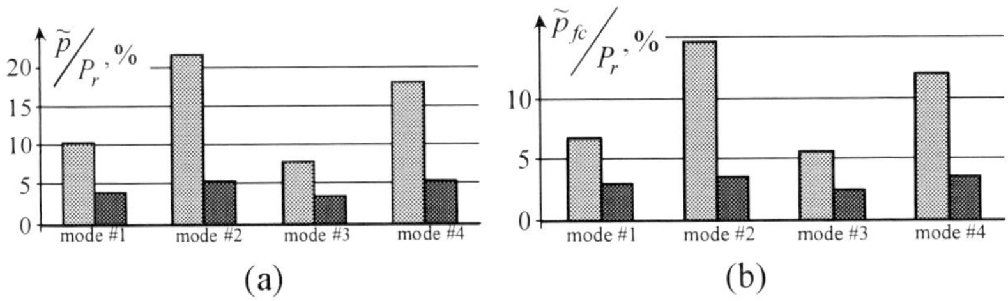

Figure 4.47. The variable components of active power consumed from the network (a) and from FC (b).

Thus, the use of the presented system of phase vector control with the adjustment to compensate for the variable component of the electromagnetic torque makes it possible to reduce the variable component of the active power consumed by an average of 68% for asymmetry in one phase and by 63% for asymmetry in two phases. In turn, the variable component of the power consumed from FC is reduced by 66% in asymmetry in one phase, and by 63% in asymmetry in two phases.

The results of the complex research confirm that at the decrease in flux linkage in the motor asymmetric phase in proportion to the asymmetry of active resistances on IM phases, a variable component of the electromagnetic torque reduces to an admissible level.

4.3.4. The Adjustment of the Phase Control System to Compensate for the Variable Component of IM Power Consumption

In the mode of compensation for the variable component of the power consumption of the asymmetric IM, the optimal value of the flux linkage of the asymmetric phase is determined according to expression [211]:

$$\varepsilon_\Psi = \varepsilon_W + \frac{1-\varepsilon_W^2}{U_{t\max}}, \tag{4.27}$$

where $U_{t\,max}$ – the maximum value of control voltage (10 V).

The research of the influence of the variation of the value of flux linkage of the asymmetric phase of IM with a power of 5.5 kW on the values of the variable components of the power consumed from the network and from FC is given in Figure 4.48.

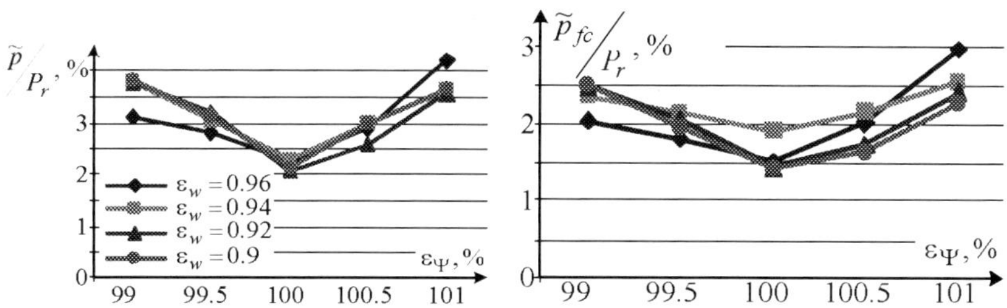

(a) (b)

Figure 4.48. The variable components of active power consumed from the network (a) and from FC (b) with variation of the flux linkage reference value.

In turn, in the considered range of changes in the reference value of the flux linkage of IM asymmetric phase, the variable component of the electromagnetic torque decreases (Figure 4.49).

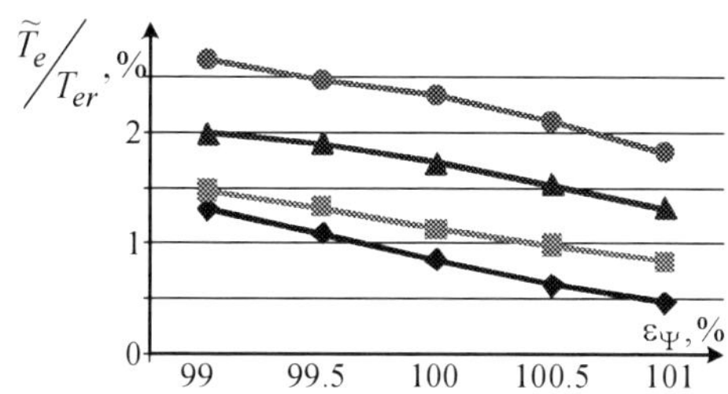

Figure 4.49. The changes in the components of IM electromagnetic torque with variations in the reference value of flux linkage.

The hodographs of the stator voltage and current and rotor flux linkage during the adjustment of the phase vector control system to compensate for the variable component of power consumption are shown in Figure 4.50.

Thus, during the adjustment of the control system to compensate for the variable component of power consumption, an asymmetric flux linkage of the motor is also formed, and the asymmetry is less than when adjusting the control system to compensate for the variable component of the electromagnetic torque. The preformed research reveals that when the flux linkage is reduced in the asymmetric phase according to expression 4.28, there is a decrease in the variable components of the active power consumed from the network and from FC to the allowable level. The redistribution of losses in the stator copper by IM phases in this case is shown in Figure 4.51.

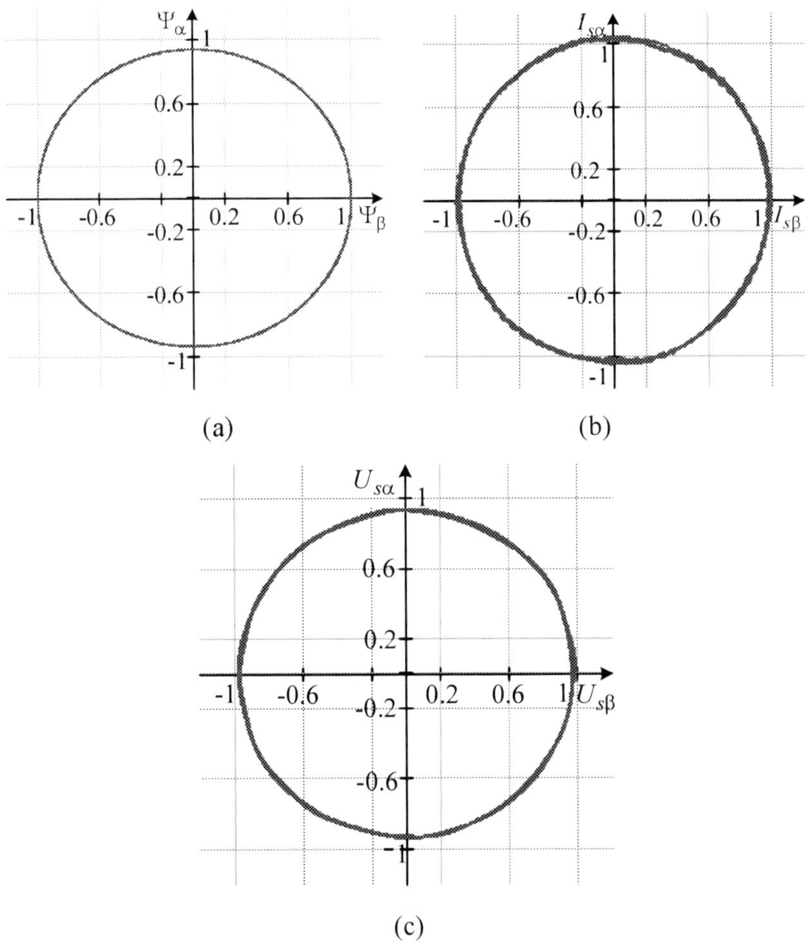

Figure 4.50. The hodographs of voltage (a), current (b) and flux linkage of the rotor (c) of IM with asymmetry in phase A equal to 10% during the adjustment of a phase vector control system to compensate for the variable component of power consumption.

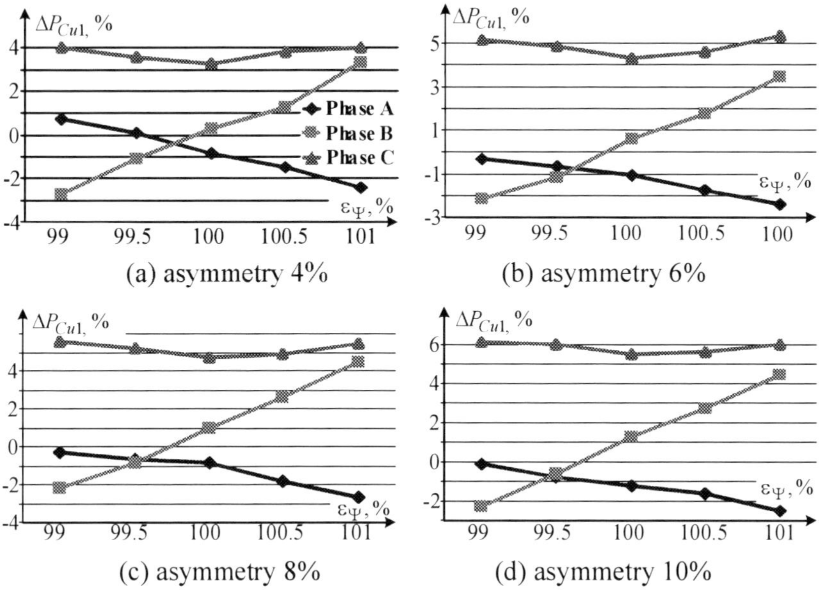

Figure 4.51. The losses in the stator copper on the phases of asymmetric IM at the variation of the reference value of flux linkage.

The results of the research also showed that when reducing the flux linkage in IM asymmetric phase in accordance with expression 4.28, there is a minimum excess of losses of the most overloaded phase of the motor stator.

The research of the modes of operation of the developed control system was conducted for these cases of asymmetry of the stator windings. Thus, the signals of active power consumed from the network and their spectral composition for the 4th mode are shown in Figures 4.52-4.53.

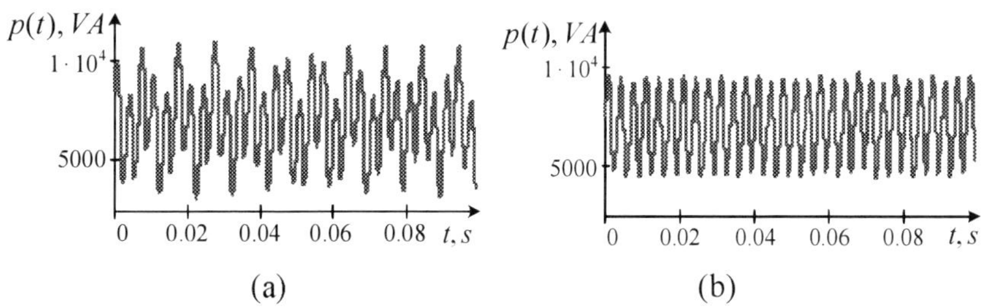

Figure 4.52. The active power consumed from the network by asymmetric IM before (a) and after (b) compensation.

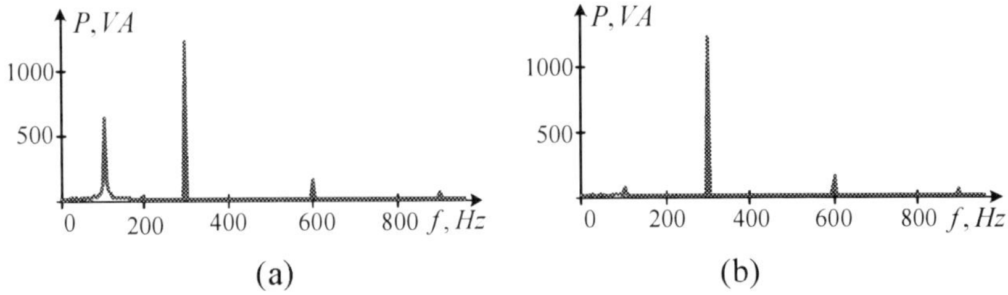

Figure 4.53. The spectral composition of the active power consumed from the network by asymmetric IM before (a) and after (b) compensation.

Thus, during the operation of the phase vector control system, there is a significant reduction of the variable component of the active power consumed from the network. The network current signals for this case are given in Figure 4.54.

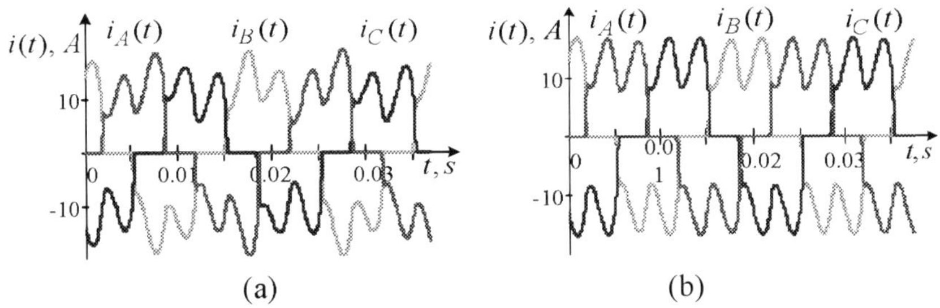

Figure 4.54. The network current signals during the operation of an asymmetric IM before (a) and after (b) compensation.

The signals of the active power consumed from FC and their spectral composition for the 4th mode of operation are given in Figures 4.55-4.56.

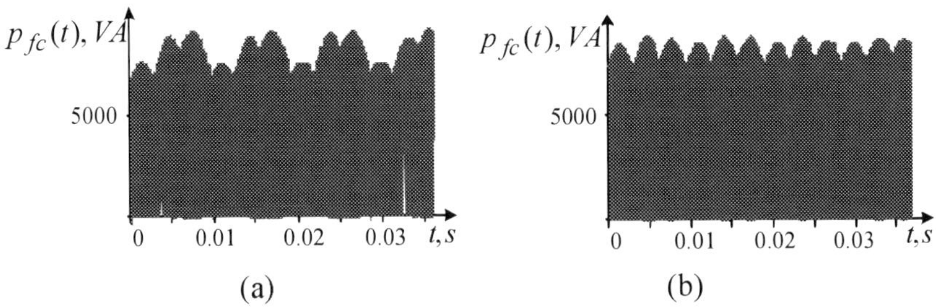

Figure 4.55. The active power of asymmetric IM consumed from FC before (a) and after (b) compensation.

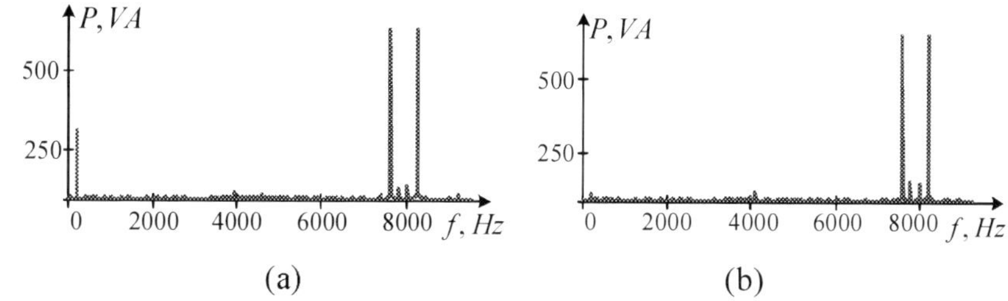

Figure 4.56. The spectral composition of the asymmetric IM active power consumed from FC before (a) and after (b) compensation.

The performed research has shown that the variable component of the active power consumed from FC is also significantly reduced. The stator current signals of the researched motor are given in Figure 4.57.

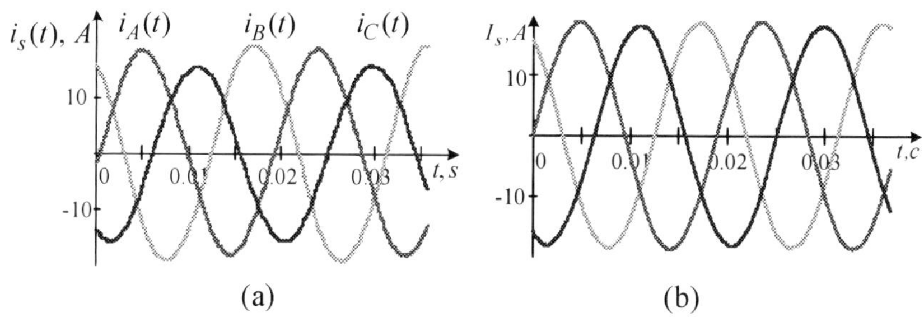

Figure 4.57. The stator current of asymmetric IM before (a) and after (b) compensation.

A comparison of the root mean square values of the variable components of the power consumed from the network and from FC for these cases of asymmetry of the stator windings of the motor is shown in Figure 4.58.

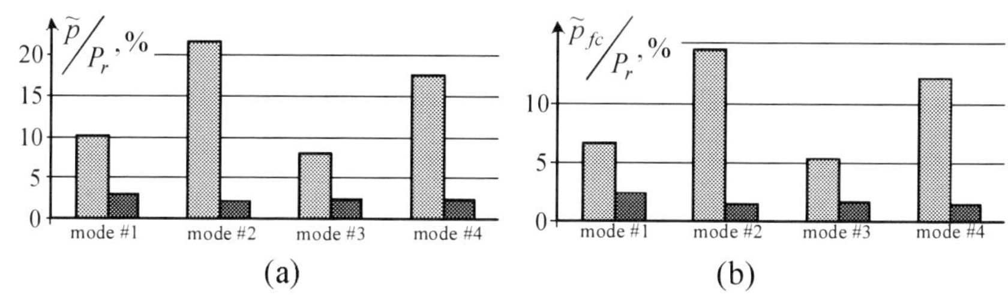

Figure 4.58. The variable components of active power consumed from the network (a) and from FC (b).

Thus, the use of the presented system of phase vector control with the adjustment to compensate for the variable component of power consumption makes it possible to reduce the variable component of active power consumption by 81% on average with the asymmetry in one phase, and by 78% with the asymmetry in two phases. In turn, the variable component of the power consumed from FC is reduced by 78% in both the case of asymmetry in one and in two phases of the motor.

The conducted research shows that in the mode of compensation for a variable component of the consumed power the variable component of IM electromagnetic torque is also compensated for (Figures 4.59-4.60).

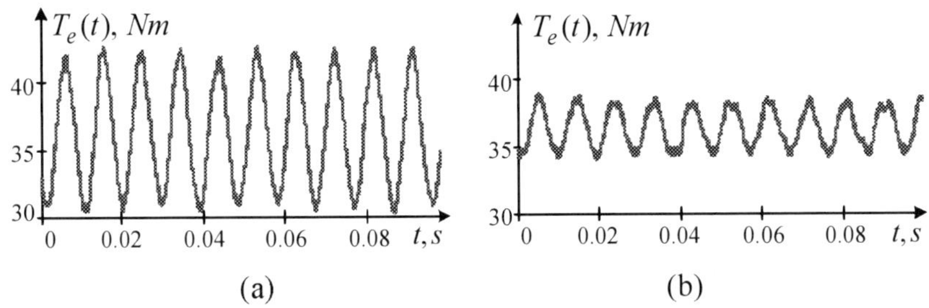

Figure 4.59. The electromagnetic torque of an asymmetric IM in the classical (a) and in the proposed (b) control system.

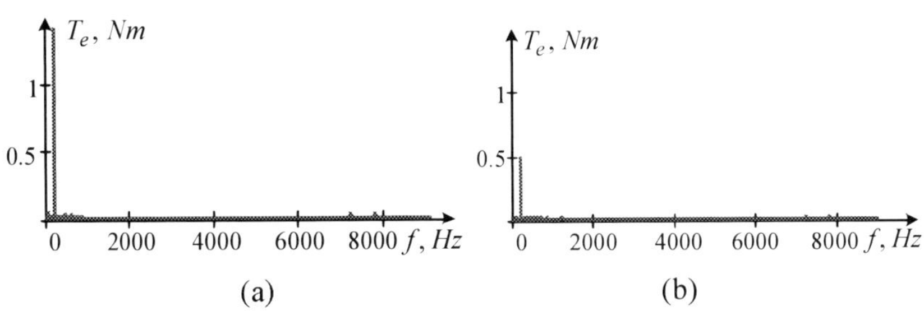

Figure 4.60. The spectral composition of the electromagnetic torque of IM before (a) and after (b) the compensation.

The results of the research show that the variable component of the electromagnetic torque in this case is reduced by 66% for cases of asymmetry in one and two phases of the motor (Figure 4.61).

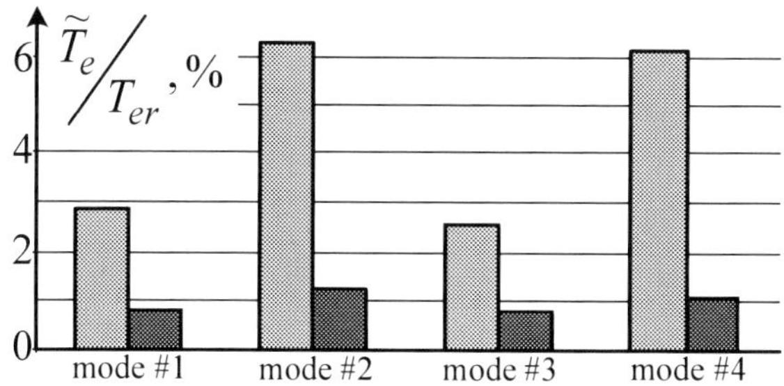

Figure 4.61. Variable components of IM electromagnetic torque.

The redistribution of losses in IM stator windings and in the power part of the ED with a phase vector control system before and after compensation is shown in Figures 4.62-4.63.

1st mode (asymmetry in phase A is 5%)

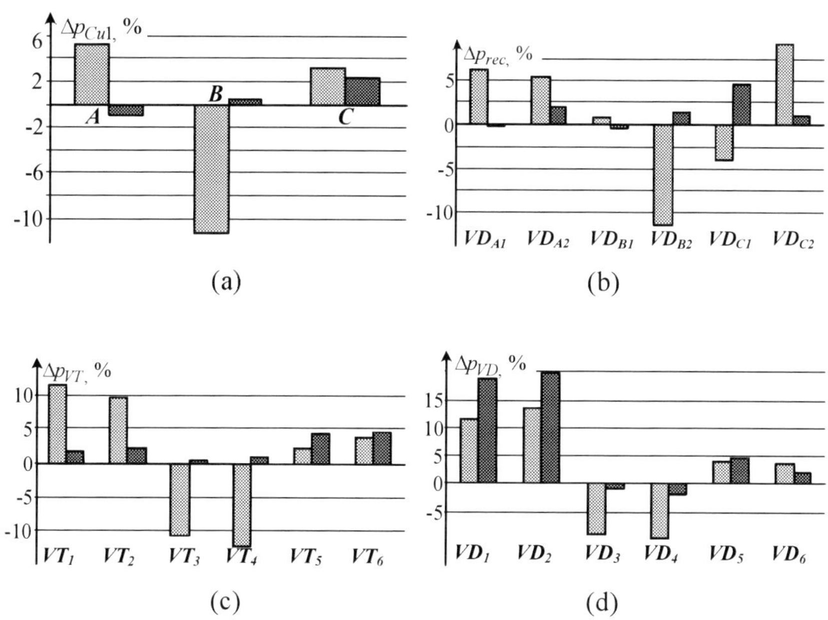

Figure 4.62. (Continued).

2nd mode (asymmetry in phase A is 10%)

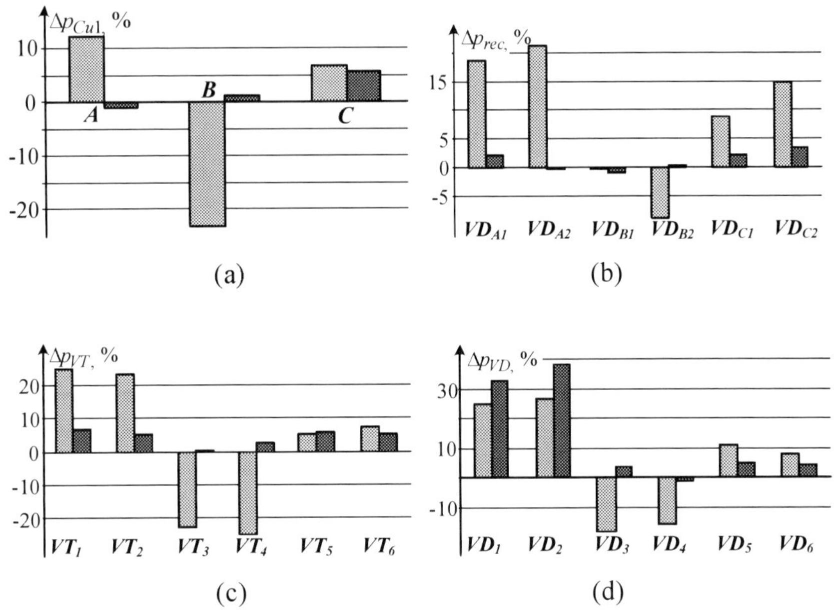

Figure 4.62. The deviation of losses in stator windings (a), input rectifier diodes (b), autonomous inverter transistors (c) and inverse diodes of transistors (d) from their rated values before and after the compensation.

3rd (asymmetry in phase A is 5%, in phase C – 3%)

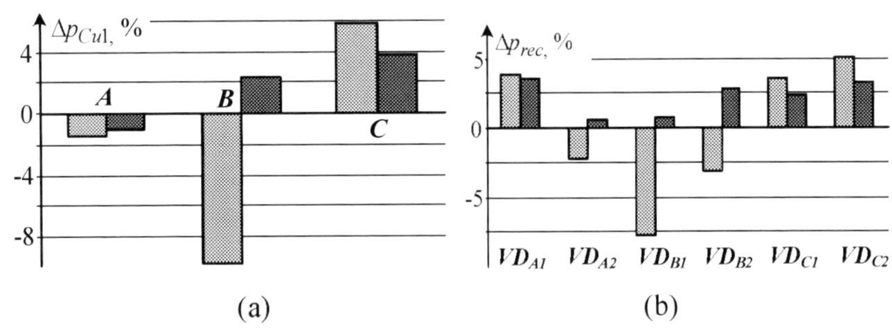

Figure 4.63. (Continued).

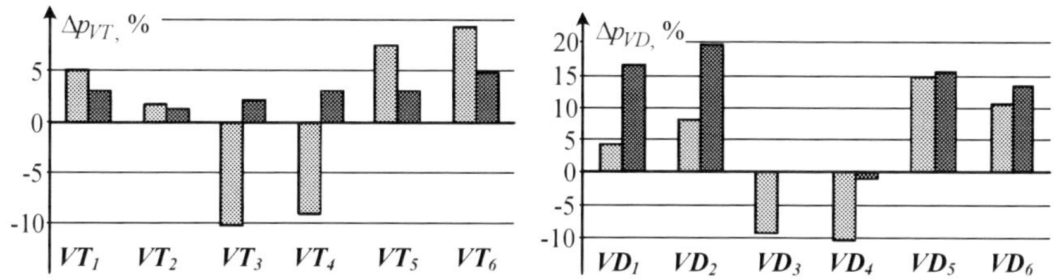

c) d)

4th mode (asymmetry in phase A is 10%, in phase C – 7%)

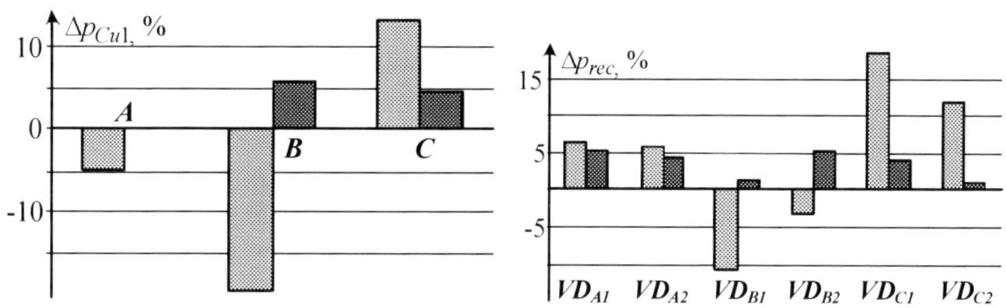

(a) (b)

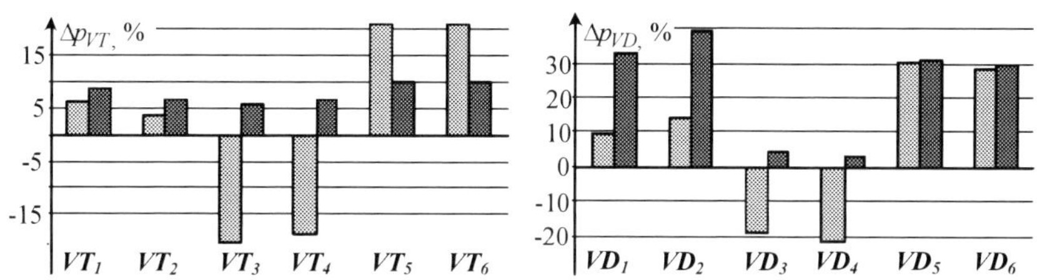

(c) (d)

Figure 4.63. The deviation of losses in stator windings (a), input rectifier diodes (b), autonomous inverter transistors (c) and inverse diodes of transistors (d) from their rated values before and after the compensation.

The research of 5.5 kW IM operation modes has shown that using the presented phase vector control system with the adjustment to compensate for the variable component of power consumption, the losses in the stator copper of the most loaded phase can be reduced by an average of 56% with the asymmetry in one phase and by 46% with the asymmetry in two phases. The losses in the power diodes of the input rectifier of the busiest phase are reduced by an average of 73% with the asymmetry in one phase, and by 61% with the asymmetry in two phases; the losses in transistor switches are reduced by an average of 66% with the asymmetry in one phase, and by 47% with the asymmetry in two phases. However, there is an increase in losses in the reverse diodes of power transistors by 31% with the asymmetry in one phase, and by 23% with the asymmetry in two phases of IM.

Thus, when reducing the flux linkage in the asymmetric phase of the motor in accordance with expression 4.28 there is a decrease in the variable components of the power consumed from the network and from FC to the allowable level.

The research of the modes of operation of the developed systems of variable-frequency electric drive with vector control has shown the possibility of reducing the variable components of the electromagnetic torque and the power consumed by an asymmetric induction motor. It is shown that to reduce the thermal overload of the individual phases of the motor and power switches of the frequency converter, the best results are shown by electric drive systems with adjustment of control systems to compensate for the variable power component. Taking this into account, the compensation strategy should be chosen in view of the requirements for the operation of specific electric drive systems of working mechanisms. The indicators of the compensation for the developed control systems are given in Table 4.5.

Table 4.5. The comparison of the indicators of compensation for the developed control systems

Damage	**Indicator decrease, %**					
	$\tilde{T}_e / T_{er}$	$\tilde{p}_e / P_r$	$\tilde{p}_{fc} / P_r$	Δp_{Cu1}	Δp_{rec}	Δp_{VT}
FTC adjusted to compensate for the electromagnetic torque variable component						
1 phase	73	42	41	38	46	29
2 phases	75			27	25	30
FTC adjusted to compensate for the active power consumption variable component						
1 phase	–	75	74	51	89	45
2 phases	–				72	39
FTC with separate regulation of IM phases adjusted to compensate for electromagnetic torque variable component						
1 phase	83	68	66	40	46	62
2 phases	77	63	63	13	27	36
FTC with separate regulation of IM phases adjusted to compensate for the active power consumption variable component						
1 phase	66	81	78	56	73	66
2 phases		78		46	61	47

Conclusion

1. Methods of compensation for the induction motor stator windings asymmetry influence on the characteristics of the systems of the variable-frequency electric drive with vector control, which are based on introduction of additional signals of compensation in closed contours of regulation of current of a stator of the motor, are developed.
2. According to the analysis of the orthogonal components of the harmonics of the electromagnetic torque of the motor in the vector control system in case of asymmetry of currents, the possibility of their compensation by correction of flux linkage in the corresponding phase is proved.
3. The system of vector control of the electric drive with separate regulation of the circuits of flux linkage and active component of current is developed individually for each phase of the induction motor. This makes it possible to adjust the phase flux linkages in accordance with the asymmetry of the electromagnetic parameters of the corresponding phase of the motor and the correction of the modes of operation of the electric drive by changing the reference signals of the phase flux linkages.
4. The possibility of using classical approaches to the synthesis of regulators of the vector control system in the regulation of periodic signals is proved. It is shown that the errors of regulation of the amplitude and phase of the signals in the closed circuits of the regulation of current and flux linkage depend on the frequency of modulation of the power switches of the autonomous voltage inverter.
5. A method for correcting the operation of induction motors with damage in the stator power circuit is developed. According to it, decreasing the flux linkage of the damaged phase reduces thermal overload of individual motor phases and semiconductor switches of the frequency converter, which extends the life of manufacturing equipment.

Chapter 5

The Determination of the Parameters of Induction Motors during Operation with a Frequency Converter

5.1. The Analysis of the Methods for the Identification of the Electromagnetic Parameters of the Induction Motor in the Start-up Period

A characteristic feature of modern digital variable-frequency electric drive consists in the presence of the function of automatic adjustment of the control system parameters for a specific object. Thus, for the calculation of the system of the vector control of the parameters and coefficients of the regulators, it is necessary to have information on the stator and rotor windings resistances (R_s, R_r), magnetizing circuit inductance (L_μ), equivalent scattering inductance of the stator circuit ($L_{s\sigma}$) and the parameters of the mechanical part of the drive. At the same time, according to research [212], the quality of control is most significantly affected by the errors in determining the resistance of the stator, the time constant of the rotor and the equivalent inductance of IM scattering. The errors in determining the stator resistance and the equivalent scattering inductance result in the fluctuations in ED coordinates. In turn, the deviation of the rotor time constant is manifested in the occurrence of a static error in calculating the angular rotation frequency, and as the error in the value of the rotor time constant increases, the error in the assessment of the slip frequency, which determines the rotor speed, grows.

The research on the sensitivity of the vector control systems to the changes in the electromagnetic parameters of IM equivalent circuit presented in [213] reveals that:

- *the change in the stator resistance* has the most significant effect on both static and dynamic characteristics of ED system, and the sensitivity to R_s change is a function of the angular rotation frequency and load. The characteristics of the drive in the low frequencies domain are most strongly influenced, up to the loss of working capacity, which is manifested in the emergence of a self-oscillating mode of operation with big pulsations of the variables;

- *the change in rotor resistance* affects the static error in the angular rotation frequency and does not affect the accuracy of the orientation of the system on the vector of flux linkage of the rotor and the dynamic characteristics of the drive. The static error in the angular rotation frequency is a function of load and does not depend on the level of the angular rotation frequency;
- *the change in mutual inductance* slightly affects the orientation by the rotor flux linkage vector and ED dynamic characteristics when working at angular frequencies lower than the rated ones. The static error in the angular frequency grows with increasing load and, when operating with a constant rotor flux linkage, does not depend on the level of the angular rotation frequency. The sensitivity of the drive to the change of L_μ significantly increases at an angular frequency exceeding the rated one up to the refusal to accept the task at an angular rotation frequency;
- *the change in scattering inductances* is noticeably manifested at stator current multiplicities exceeding 2-3 of the rated value, and it usually does not exceed 30% reduction relative to the unsaturated value. Even 50% change of $L_{s\sigma}$ does not significantly affect ED characteristics.

Thus, to provide ED steady operation and significantly expand the range of the rotation frequency control with constant torque, it is necessary to have adequate information about the electromagnetic parameters of the motor, which may change during operation of the electric drive depending on time, external influences and ED coordinates.

In the general case, the algorithms for identifying the parameters of IM equivalent circuits can be grouped into the algorithms for preliminary and current identification, active and passive ones [214]. The preliminary (*"Off-line"*) identification of parameters is carried out in advance in preparation of ED for the operation. When this identification is implemented, a rotating magnetic field is not created in IM deliberately. The parameter assessments obtained during the preliminary identification process are used when setting up the self-tuning control system, and are also used as initial approximations for the current identification algorithms. In turn, the current *("On-line")* identification is carried out during the operation of the electric drive and is used to monitor changes in IM parameters due to changes in the motor temperature and operating point on the magnetization curve. The current identification is more complex due to its structure, although its implementation determines fewer parameters. This mainly results from the fact that some of the parameters have already been determined in the process of preliminary identification and are considered unchanged. In addition, during ED operation, the control is carried out through two channels (the flux and the torque), and it is much more difficult to link the parameters of transients with the parameters of the control object, taking into account the operation of the regulators.

Active identification implies the use of special test effects, passive one, on the contrary, is based only on the measurement of work processes by electrical variables and in estimating the coordinates of the state – by the variables determined by identifiers. Active algorithms are the best for the tasks of preliminary identification of ED parameters, as the course of the technological process is not disturbed. It is possible to focus on the impact on a particular parameter with the help of special test signals. Any interference in the technological process is undesirable during the current identification of the use of the test signals.

As a rule, the function of automatic adjustment of parameters of the control system is implemented by means of a set of test modes [215, 216]. All the parameters of the variable-frequency ED are initially set during these test modes. In the mode of ED automatic adjustment the control system determines all the necessary parameters based on the information on the rated motor voltage, rated stator current, rated frequency, rated slip, rated power factor, the number of pole pairs.

The effectiveness of the automatic parameter adjustment consists in the simplicity and ability to implement auto-adjustment modes exclusively by ED internal resources in the real conditions of its operation. A number of known methods of experimental determination of the motor parameters [183] require additional operations associated, for example, with the fixation of the shaft in the mode of IM short circuit, rupture of the kinematic connection of the motor with the mechanism in the idle mode, measuring the stator voltage in the mode of its disconnection from the DC power supply. Therefore, special algorithms are required that do not affect the course of the technological process, do not need any manipulation of the mechanism or the use of additional devices. Thus, to identify IM electromagnetic parameters in the variable-frequency ED, it is necessary to synthesize such test effects, in which the motor would not induce a rotating magnetic field, but additionally take into account the following factors [214]:

- in the absence of the rotor rotation and the rotating magnetic field there is no EMF in IM stator, which is a significant proportion of the rated voltage. Therefore, the preliminary identification mode uses low voltage levels, which makes high demands on the accuracy of FC control system or phase voltage sensors;
- if ED does not provide for the regulation of the magnetic flux, at the stage of the preliminary identification in determining the inductances, all measurements must be performed at the values of magnetizing current that are close to the rated ones. In the case when it is planned to change IM magnetic flux, it is necessary to perform the procedure for determining the inductances at different values of the magnetizing current. In both cases, you need to perform several repetitions of the same procedure.
- it is necessary to take into account the nature of the output voltage of the converter and to measure the current at certain points in time, synchronized with the PWM reference signal;
- additional electrical elements are included in the stator circuit of the motor, such as inverter power switches with nonlinear resistance and motor power cable, which is characterized by distributed active and reactive resistances at high frequencies.

Taking into account these factors in the auto-adjustment mode of ED system, the simplest test modes include the modes when the motor is powered by a “fixed” voltage vector (frequency of rotation of the voltage vector is zero), as well as the real idle mode.

In the modern systems of variable-frequency ED the following approaches are used for the problem of the identification of induction motor internal parameters.

5.1.1. The Stator Resistance

It is determined in the "fixed" vector mode [217], according to the formation of a constant equivalent voltage on the stator windings. The operations are performed in the following sequence:

- the level of direct voltage corresponding to the rated current of the motor is determined in the mode of step-by-step setting of the output voltage and control of the current amplitude at each step;
- a series of measurements of the stator current at the set DC voltage level is performed;
- the resistance of the stator circuit is calculated by expression:

$$R_s = U_s / I_{s\,mean}, \tag{5.1}$$

where U_s – the constant component of the stator voltage; $I_{s\,mean} = \frac{1}{n}\sum_{i=1}^{n} I_{si}$ – average current, n – the number of measurements of the stator current.

5.1.2. The Stator Inductive Reactance

5.1.2.1. Method 1 [218]

The value of the equivalent inductance of the stator scattering can be determined in the "fixed vector" mode, for which two samples of the stator current (I_{s1}, I_{s2}) are taken at the interval of turning on the zero voltage vector within the modulation cycle (Figure 5.1).

The value of σL_s is calculated on the basis of the linear approximation of the curve of the change of current instantaneous value by expression:

$$\sigma L_s = \frac{1}{n}\sum_{i=1}^{n} \frac{R_s I_{s\,mean} \Delta T}{(I_{s1})_i - (I_{s2})_i}, \tag{5.2}$$

where σ – scattering coefficient; ΔT – the interval between current samples.

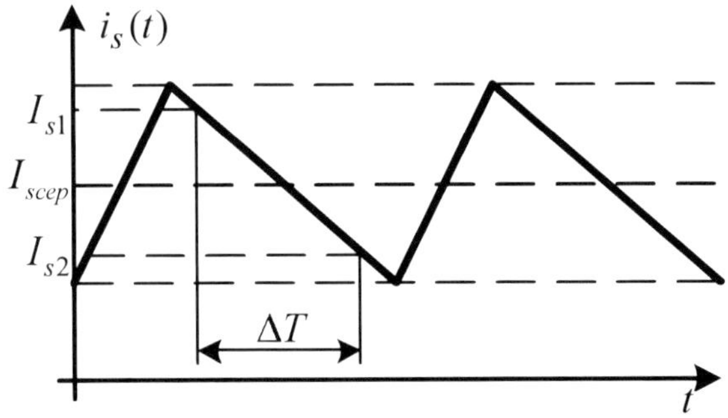

Figure 5.1. The determination of the equivalent scattering inductance of the stator circuit.

5.1.2.2. Method 2 [214]

A constant voltage is applied to the stator, and the stator current is measured in a steady state. Then the voltage of the converter is measured at the same value of the current, but provided that the current and voltage vary from zero according to the linear law. The difference in voltages characterizes the total inductance of the stator. When the stator inductance is determined in this way, the converter is to meet high requirements as to the reproduction of the set voltage, because the range of voltage changes is quite small, and the time of linear voltage rise in this range is quite large (up to several seconds).

5.1.2.3. Method 3 [219]

The motor magnetization mode is determined by maintaining a current close to the rated value, after which the system is switched to the mode of maintaining the specified value of the flux linkage. The error in determining the flux linkage during the identification test will not have time to accumulate significantly under the condition of predetermined stator resistance. Based on this, the inductance of the stator is calculated by expression:

$$L_S = \Psi_S / I_{S\ mean}, \tag{5.3}$$

where $\Psi_S = \int (U_S - I_S R_S) dt$ – the stator flux linkage.

5.1.2.4. Method 4 [161]

In the steady state of the stopped electric drive, when applying uniform unipolar modulation and the formation of a "fixed vector" of the output voltage (Figure 5.2), the total scattering inductance can be determined according to the expression:

$$L_\sigma = u_s \Big/ \frac{di_s}{dt}, \tag{5.4}$$

where u_s – the stator voltage modulus; di_s/dt – the derivative of the stator current modulus on the *i*-th measurement interval carried out on the leading time interval of the output voltage of the converter (indicated as τ_1 in Figure 5.2); $L_\sigma = L_s - L_\mu^2 / L_r = L_{\sigma s} + kL_{\sigma r}$.

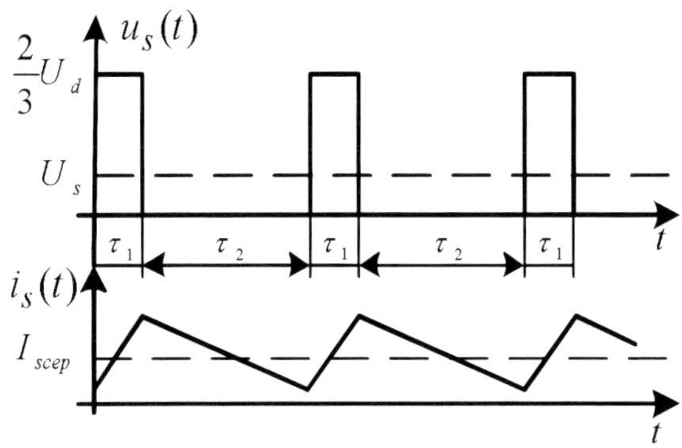

Figure 5.2. The time diagrams of the moduli of the generalized voltage and current vectors of IM stator when powered by SVI with PWM.

5.1.2.5. Method 5 [220]

Low voltage at high output frequency of the frequency converter is applied to the stator windings of the motor. The value of inductance is calculated based on the expression:

$$L_s = \frac{U_s I_x}{2\pi f_u I_{s\,mean}^2}, \tag{5.5}$$

where f_u – the frequency of the output voltage of the converter; I_x – the reactive component of the stator current.

5.1.3. The Inductance of the Magnetization Circuit

5.1.3.1. Method 1 [217]

It is determined in idle mode at a frequency close to the rated motor frequency. The stator voltage is formed in accordance with the law – $U_s / f_u = U_n / f_n$. The motor accelerates to the specified frequency ($f_u \approx 0.9 f_b$), and a series of *n* stator current measurements is performed. The value of the parameter is calculated by expression:

$$L_{\mu} = \frac{1}{\omega_s} \frac{E_{mean}}{I_{\mu\, mean}}, \tag{5.6}$$

where E_{mean}, $I_{\mu\, mean}$ – the average values of EMF and magnetizing current on the *i*-th interval of measurements; ω_s – the field rotation frequency.

5.1.3.2. Method 2 [220]

The rated output voltage with high frequency is applied to IM stator windings. FC output voltage approximately corresponds to the natural characteristic $U / f = const$. In this case, the calculation is based on expression:

$$L_{\mu} = \frac{U_s}{I_x} - L_s . \tag{5.7}$$

5.1.3.3. Method 3 [161]

In IM steady operating mode, at an angular rotation frequency not equal to zero, the value of the reduced magnetization inductance is determined on the basis of expression:

$$kL_{\mu} = \frac{kE_r}{\omega I_1}, \tag{5.8}$$

where ω – the frequency of the stator current fundamental harmonic; I_1 – the fundamental harmonic of the stator current; kE_r – the modulus of the generalized EMF vector of the motor rotor reduced to the stator, calculated from the following dependences:

$$\begin{cases} kE_r = \sqrt{\frac{4}{3}\left((kE_{ra})^2 + kE_{rz}kE_{rb} + (kE_{rb})^2\right)}; \\ kE_{ra} = U_{sa} - R_s I_{sa} - L_\sigma \frac{dI_{sa}}{dt}; \\ kE_{rb} = U_{sb} - R_s I_{sb} - L_\sigma \frac{dI_{sb}}{dt}, \end{cases}$$

where U_{sa}, U_{sb}, I_{sa}, I_{sb} – two present values of the voltage and current of IM stator.

5.1.4. Rotor Resistance

5.1.4.1. Method 1 [221]

The calculation is based on the IM equivalent circuit by expression:

$$R_r = \frac{E_{rr} s_r}{I_{rr}}, \tag{5.9}$$

where $I_{rr} = \sqrt{I_s^2 - I_\mu^2}$, ($I_{\mu r} = E_{rr}/\omega_s L_\mu$) – the calculated values of the rotor current and magnetization for the rated mode of motor operation; E_{rr} – the calculated value of the EMF of the rotor for the rated operating mode; ω_s – the rated frequency of the rotor rotation; s_r – the rated slip of the motor.

5.1.4.2. Method 2 [222]

The test mode of single-phase power supply of IM with alternating voltage of the lowered frequency is used, thus power is supplied on two IM phases, and the third phase is disconnected. In this case, the motor shaft is stationary, and accordingly there is no need for its mechanical fixation. The introduction of a small constant component into the supply voltage provides additional stabilization of the shaft in a stationary state. The processes in the phase relative to the variable component are equivalent to the processes of IM short-circuit mode at low power frequency, which is implemented by means of mechanical fixation of the rotor shaft. The frequency of the alternating component of the stator voltage is taken from ratio $\omega_{sc} \approx R_s / L_\mu$, which makes it possible to minimize the calculated error caused by the errors of variables measurement and assumptions accepted in calculations. Besides, there is practically no effect of displacement of current in rotor conductors. The rotor active resistance is assessed on the basis of the analysis of IM simplified equivalent circuit with neglect of inductances of scattering of a stator and a rotor ($R_s, R_r, \omega_{sc} L_\mu >> \omega_{sc} L_{\sigma s}, \omega_{sc} L_{\sigma r}$) by the following equations:

$$R_r = \frac{E_\mu}{\sqrt{I_s^2 - I_\mu^2}}; \qquad (5.10)$$

where $E_\mu = \sqrt{u_s^2 - 2u_s R_s i_s \cos\varphi_{sc} + (R_s i_s)^2}$; $I_\mu = E_\mu / \omega_{sc} L_\mu$ – the calculated values of EMF and magnetization circuit current; u_s, i_s – the current values of the fundamental harmonics of the variable components of the stator voltage and current; φ_{cs} – the angle between the fundamental harmonics of the voltage and current of IM phase.

The use of the set value of the output voltage of the inverter is allowed as the fundamental harmonic of the stator phase voltage. In order to reduce the calculation error related to the inaccurate reproduction of the stator voltage of a set value, the experiment is performed at a low modulation frequency (0.5-2 kHz).

5.1.4.3. Method 3 [222]

The test mode of switching on the "fixed" voltage vector at IM zero initial conditions is used. The current curve records a constant value of current (I_y) and the value of current at the breaking point (I_u of fast processes and the transition to slow ones, which are characterized by a constant time of the rotor (Figure 5.3).

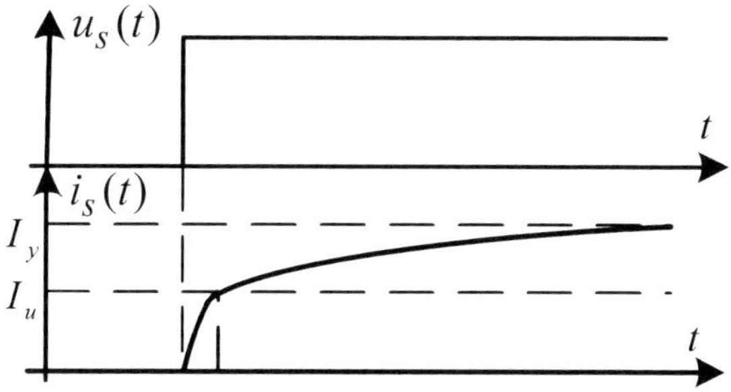

Figure 5.3. The time diagrams of the test mode for the determination of the rotor resistance.

Based on this, the rotor resistance is determined by the following expression:

$$R_r = R_s(I_y - I_u)/I_u . \quad (5.11)$$

The accuracy of this method is largely determined by the accuracy of fixing the breaking point of the current curve, which is provided by the control system in automatic mode based on the mathematical analysis of the measurement samples.

The equivalent scattering inductance of the stator circuit and the inductance of the magnetization circuit can be calculated in the mode of switching on the "fixed" voltage vector at IM zero initial conditions, in addition to the rotor resistance, as the current curve includes components of both fast (stator) and slow (rotor) dynamics. However, they can be distinguished in the pure form (regardless of other unknown parameters) only based on the assumptions that affect the accuracy of the estimates L_μ and σL_s .

5.1.4.4. Method 4 [219]

The magnetization mode is used; a linear approximation of the rotor magnetization curve is performed. The value of resistance is calculated by expression:

$$R_r = -\Psi_r/t_f I_{r\,mean} , \quad (5.12)$$

where Ψ_{r} – the value of the flux linkage of the rotor at the time of transition to the mode of maintaining the specified value of the flux linkage; t_f – magnetization time with current limitation; $I_{r\ mean}$ – the average value of the component of the rotor current over time t_f.

5.1.4.5. Method 5 [161]

In the steady state of IM fixed shaft, when applying the uniform unipolar modulation and the formation of a "fixed vector" of the output voltage, the value of the motor rotor resistance reduced to the stator is calculated based on expression:

$$k^2 R_r = R_\Sigma - R_s\,, \tag{5.13}$$

where R_Σ – the total IM resistance calculated based on expression:

$$R_\Sigma = \left(U_s + L_\sigma \frac{di_s}{dt}\right) \Big/ \left(I_s - i_s\right),$$

where i_s, di_s/dt – the stator current and its derivative on the *i*-th measurement interval carried out on the non-conductive time interval of the output voltage of the converter (indicated as τ_2 in Figure 5.2).

5.1.4.6. Method 6 [220]

In the test mode, FC output frequency changes abruptly, while the current throw is controlled. The calculation is based on expression:

$$R_r = \frac{U_2}{f_2} \cdot \frac{f_2 - f_1}{I_{r2} - I_{r1}} = \frac{E_1}{I_{r2} - I_{r1}} \left(\frac{f_2}{f_1} - 1\right), \tag{5.14}$$

where f_1, f_2 – the frequency of the output voltage of the converter before and after the jump; I_{r1}, I_{r2} – the active component of the stator current, respectively, before and after the jump; U_2 – output voltage of the converter after the jump; E_1 – the motor EMF before the jump.

Thus, the function of automatic adjustment of the control system parameters to a real object can be implemented with the help of a set of different test modes. An example of the combined form of the stator voltage signal at the stage of preliminary identification of the parameters in the motor power supply mode with a "fixed" voltage vector [223] is shown in Figure 5.4. a, and an example with the change of the voltage and the frequency of test effects [220] is shown in Figure 5.4, b.

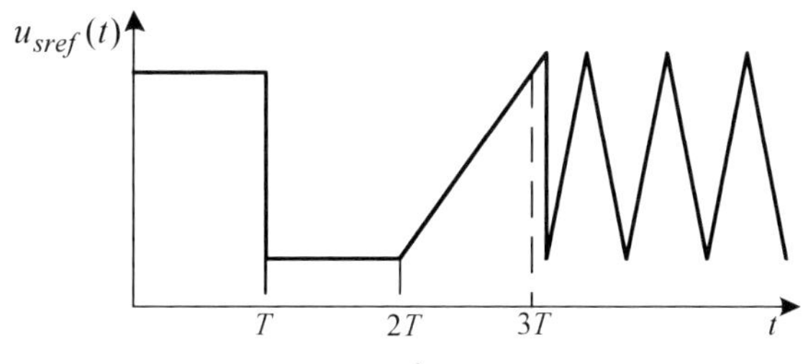

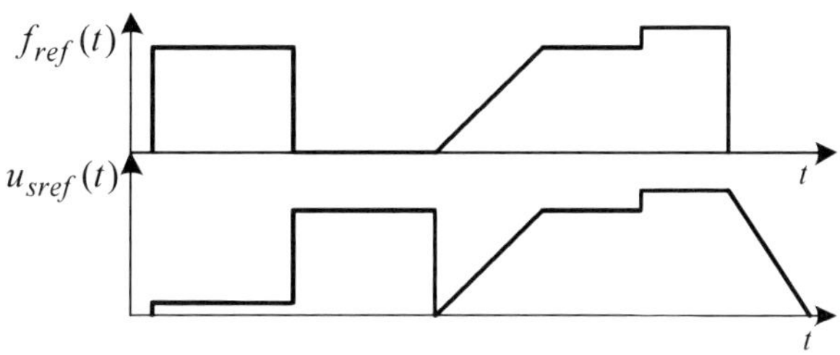

Figure 5.4. The forms of the test reference signals at the stage of the preliminary identification of the parameters.

The measurement and recording of current and voltage signals of AIR80V4U2 IM with a power of 1.5 kW supplied by a Lenze Vector 8200 FC in the pre-identification mode were performed with the help of the developed measuring module [152] (Figure 5.5).

Therefore, the methods for determining the electromagnetic parameters of IM equivalent circuit in the pre-start-up period are based on a set of different test effects on the motor stator windings with subsequent processing of the information, based on the assumption that the electrical and magnetic systems of motors are symmetrical.

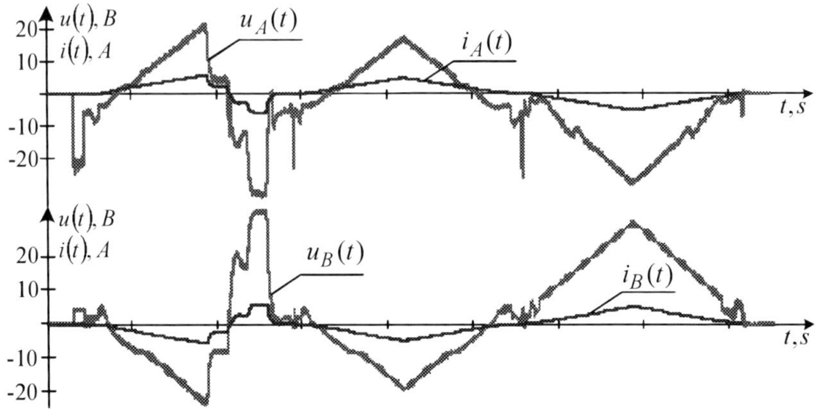

Figure 5.5. The experimental signals of IM phase voltages and currents in the mode of preliminary identification.

5.2. The Assessment of Induction Motor Parameters Based on the Low Frequency Sinusoid Test Effects

When IM is powered by a frequency converter, the motor windings are in most cases connected as a "star" without a neutral wire and two phases are involved in the operation at the same time. In this case, it is possible to use the same approaches to the construction of methods for identifying parameters as in the connection of the "star" with a neutral wire, and which characterize the electrical and magnetic properties of the motor. The only difference will be that the identification mode determines the parameters of not one phase of the motor, but two.

The problem of identifying the parameters of IM as part of the variable-frequency ED can be solved by using sinusoidal test effects on IM stator windings [224-226]. The electromagnetic and energy relations are analyzed at different frequency power supply of low frequency voltage. The setting of sinusoidal test effects with low frequency enables avoiding a number of disadvantages, such as: the low accuracy of determining the parameters of the magnetization circuit due to the fact that at high frequencies of the supply voltage the magnetizing current is less than the stator and rotor currents by a factor of ten; the need to take into account additional physical effects (current displacement effect, eddy currents, etc.) that occur on higher harmonics and are difficult to mathematically describe in conditions of limited information about the researched object; the need for mechanical fixation of IM shaft.

Signals with any frequency in the range from 0 to 50 Hz can be used to form test effects. Based on the experimentally measured signals of current and voltage of IM stator the resistive impedance and reactance are defined: $R_\Sigma = \mathrm{Re}(U_1/I_1), X_\Sigma = \mathrm{Im}(U_1/I_1)$, where U_1, I_1 – the first harmonics of the current and voltage. The determination of IM electromagnetic parameters is based on solving a system of equations consisting of equations of complex impedance determined for current and voltage signals at different low frequencies [227]. In this case two equations are written for each signal – the real and the imaginary part. Due to the fact that in the low frequency supply voltage losses in steel are insignificant, the value of the resistance of the magnetization circuit can be neglected [226, 228]. Then the system of equations for determining IM electromagnetic parameters will be of the form:

$$\begin{cases} R_{\Sigma n} = R_s + \dfrac{R_r (L_\mu \cdot \omega_n)^2}{R_r^{\,2} + (L_{r\sigma}\omega_n + L_\mu \omega_n)^2}; \\ \mathrm{X}_{\Sigma n} = L_{s\sigma}\omega_n + \\ + \dfrac{R_r^{\,2} L_\mu \omega_n + L_\mu \omega_n (L_{r\sigma}\omega_n)^2 + (L_\mu \omega_n)^2 L_{r\sigma}\omega_n}{R_r^{\,2} + (L_{r\sigma}\omega_n + L_\mu \omega_n)^2}, \end{cases} \tag{5.15}$$

where $R_{\Sigma n}, \mathrm{X}_{\Sigma n}$ – experimental total values of resistance and inductive reactance of the windings of the researched motor at different low frequencies; ω_n – the angular frequency of supply voltage.

The methods of numerical solution of systems of algebraic equations require the presence of initial approximations and the range of change of the unknown parameters [229, 230], on which the accuracy of determining the unknown parameters depends [231]. When solving the above system of equations (5.15), the conditions under which the calculations will be performed must be taken into account. In this case, the initial approximations of the required parameters of the IM equivalent circuit [232] are determined as follows:

- for the rotor resistance – by the difference of the total resistance R_Σ at a mains frequency of 50 Hz and the stator resistance: $R_r = R_\Sigma - R_s$ (R_s is determined in the mode of "fixed" vector by applying a DC voltage to the windings);
- for the stator inductive reactance – as $X_s = X_\Sigma / 2$, where X_Σ – total inductive reactance at a mains frequency of 50 Hz;
- for the rotor inductive reactance – as $X_r = X_s \cdot c$, where coefficient c, depending on the machine type, changes within the range of 1.01–1.05;
- for inductive reactance of the magnetization circuit – the difference between the imaginary part of the complex impedance calculated from the experimental signals in the short-circuit mode at low frequencies, and the inductive reactance of the stator: $X_\mu = X_\Sigma - X_s$.

The determination of the range of the change of the unknown electromagnetic parameters is based on the dependences of the real and imaginary part of the complex impedance of IM T-shaped equivalent circuit on the frequency. The range of deviations of the rotor resistance is determined from the equations of the real part of the complex impedance, and for inductive reactance – from the equations of the imaginary part of the complex impedance at the power supply fundamental frequency equal to 50 Hz. Thus, to determine the ranges of electromagnetic parameters of IM equivalent circuit, for 4A motors [173] the calculations were performed for low power motors (<5 kW), medium power (5–250 kW) and high power ones (> 250 kW), the published data of which are given in Table 5.1.

The curves of deviations of the identified values relative to the true value of the rotor resistance and the sum of the inductive reactance of the stator and rotor from the frequency of the supply voltage for the considered ranges of motor power are shown in Figure 5.6.

Based on the performed calculations, Table 5.2 contains the range of deviations of the electromagnetic parameters of IM equivalent circuit in the identification of the stator resistance and inductive reactance of the stator and rotor.

Table 5.1. The published data of 4A motors

Type	P_n, kW	n, rev/min	R_s	$X_{s\sigma}$	R_r	$X_{r\sigma}$	X_μ
Low power motors (< 5 kW)							
4AA50A2U3	0.09	3000	101.1	61.99	94.34	94.34	168.5
4AA71B6U3	0.55	1000	13.33	9.162	12.5	14.16	116.6
4A80B4U3	1.5	1500	7.38	4.799	4.245	7.38	116.9
4A100L6U3	2.2	1000	3.73	4.269	2.6	8.15	73.74
4A100S4U3	3	1500	2.56	2.588	1.736	4.26	72.07
4A100S2U3	4	3000	1.5	1.529	1	2.75	94.49
Medium power motors (5–250 kW)							
4A100L2U3	5.5	3000	1.254	1.13	0.753	2.3	79.44
4A160M6U3	15	1000	0.454	0.733	0.205	1.17	21.99
4A200L4U3	45	1500	0.09	0.218	0.045	0.37	12.22
4A280S6U3	75	1000	0.05	0.189	0.033	0.21	5.83
4A280M2U3	132	3000	0.015	0.085	0.012	0.089	3.74
4A355M8U3	160	750	0.014	0.086	0.012	0.115	2.15
4A315M4U3	200	1500	0.0087	0.054	0.0087	0.075	2.56
High power motors (> 250 kW)							
4AH355M4U3	250	1000	0.01	0.054	0.0073	0.064	1.955
4A355M2U3	315	3000	0.005	0.034	0.0043	0.043	2.367
4A355M4U3	315	1500	0.0048	0.006	0.0056	0.056	2.272
4AH355M2U3	400	3000	0.0041	0.026	0.0032	0.035	1.767
4AH355M4U3	400	1500	0.0059	0.034	0.0041	0.043	1.676

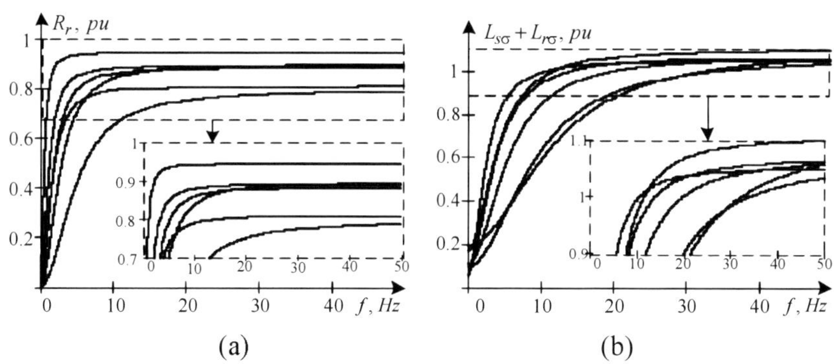

Figure 5.6. (Continued).

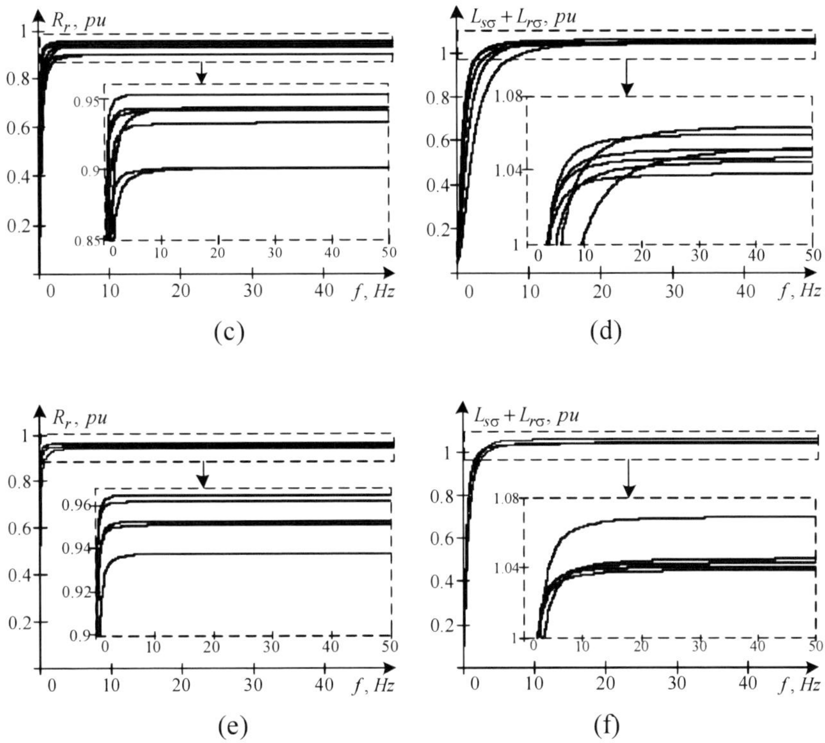

Figure 5.6. The dependences of the deviations of the identified values relative to the true values of the rotor resistance (a, c, e) and the sum of inductive stator and rotor scattering resistances (b, d, f) on the frequency for low (a, b), medium (c, d) and high power (e, f).

Table 5.2. The deviation of the identified parameters of IM equivalent circuit

Motor power	R_r, **min/max, %**	$L_{s\sigma}$, **min/max, %**	$L_{r\sigma}$, **min/max, %**
Low	5.5/21	0.15/3.2	0.2/3.3
Medium	4.5/10	0.6/1.5	0.6/1.6
High	3.5/5	0.5/1.1	0.5/1.2

Determining the range of deviations of the identified values of the inductance of the magnetization circuit is based on the consideration of the T-shaped IM equivalent circuit taking into account the slip [233] (Figure 5.7).

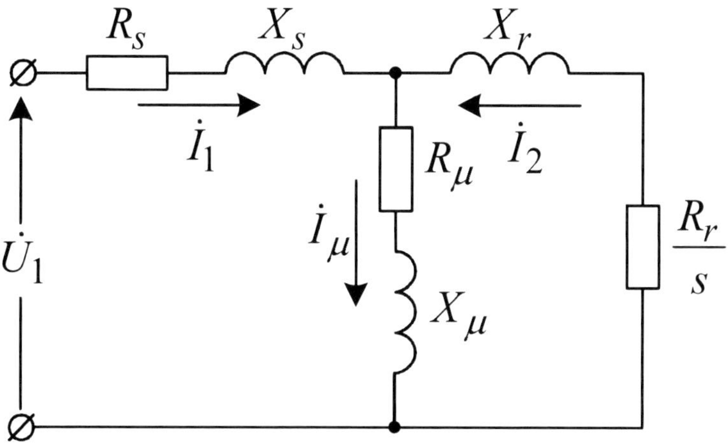

Figure 5.7. T-shaped IM equivalent circuit for rotor rotation mode.

Accordingly, the equation of the imaginary part of the complex impedance, taking into account the slid of the motor is of the form:

$$X_\Sigma(\omega, s) = L_s\omega + \\ + \frac{R_\mu{}^2 L_r\omega + (R_r/s)^2 L_\mu\omega + L_\mu L_r^2\omega^3 + L_\mu{}^2 L_r\omega^3}{\left((R_r/s) + R_\mu\right)^2 + \left(L_r\omega + L_\mu\omega\right)^2}. \quad (5.16)$$

The curves of the deviation of the identified values relative to the true value of the inductance of the magnetization circuit at different frequencies of the power supply are shown in Figure 5.8.

The graphs of the dependence of the identified values relative to the true value of the inductance of the magnetizing circuit on sliding at different frequencies of the supply voltage are shown in Figure 5.9.

The performed calculations show that the minimum deviations of the identified values of the inductance of the magnetization circuit will be obtained at small values of the supply voltage and small values of the slip. Table 5.3 contains the maximum and minimum deviations of this parameter in percentage form.

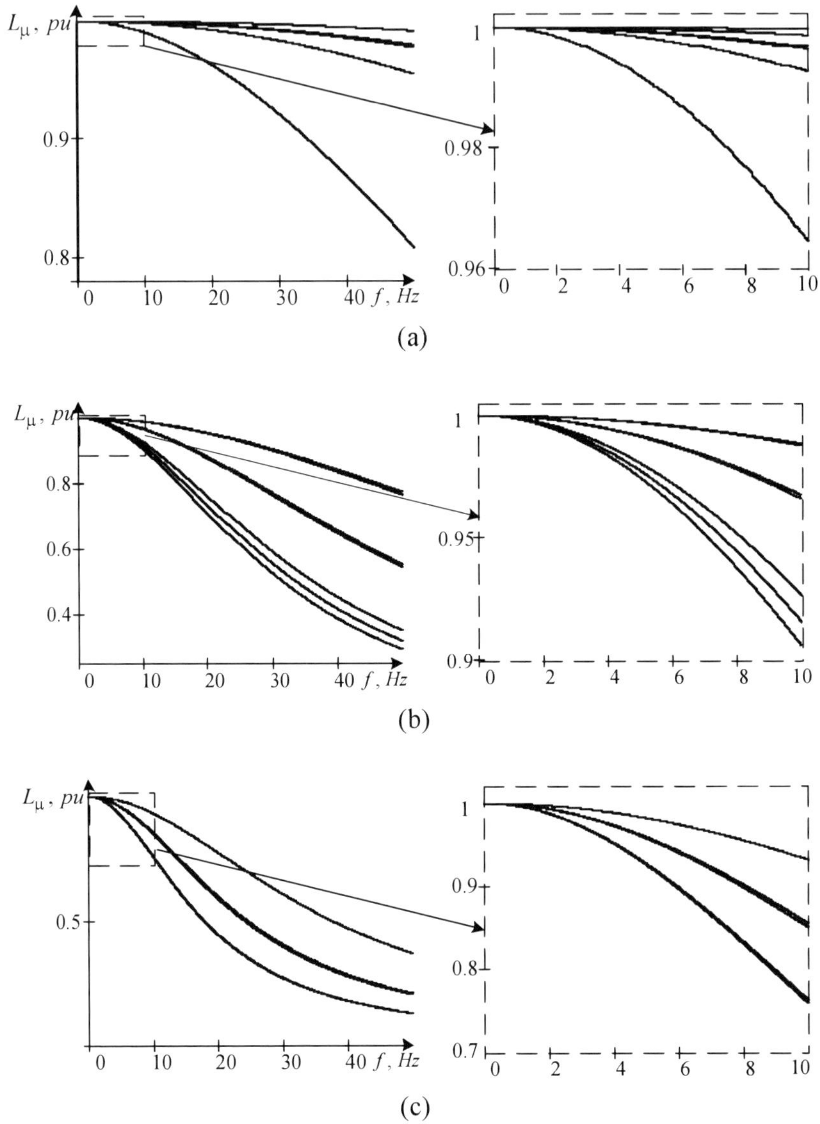

Figure 5.8. The dependence of deviations of the identified values relative to the true value of the inductance of the magnetization circuit on the supply voltage frequency for low (a), medium (b) and high power (c) Ims.

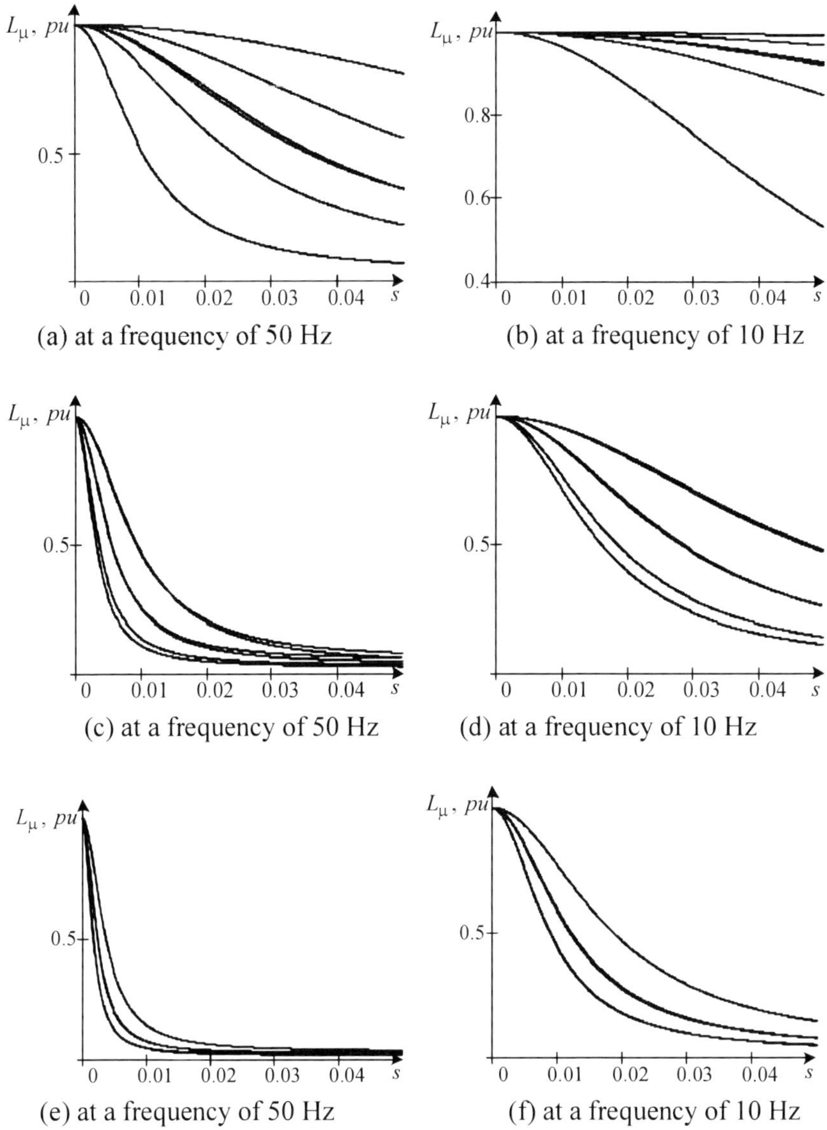

Figure 5.9. The dependence of deviations of the identified values relative to the true values of the inductance of the magnetization circuit on the motor slip for motors of low (a, b), medium (c, d) and high power (e, f) at different supply voltage frequencies.

Table 5.3. Changing the inductance of the magnetization circuit depending on the frequency of the supply voltage

Motor power	Frequency, Hz		
	50	**30**	**10**
Low	0.2/47	0.3/24	0.03/3.5
Medium	22/88	10/75	1.2/28
High	63/95	40/90	7/55

Complementing the system of equations (5.15) with initial approximations and ranges of the change of the unknown parameters, the electromagnetic parameters of IM equivalent circuit are calculated. The solution to the system of nonlinear equations is based on the use of the method of rapid descent [234, 235], in which the direction in which the functional decreases is determined at a given approximation, and the approximation is moved in this direction. If the amount of displacement is not very large, the value of the functional will definitely decrease. The use of this method is based on the consideration of the Himmelblau function, for which the points of local minimum are found.

The algorithm of the program for identification of electromagnetic parameters of IM equivalent circuit by the method of rapid descent is shown in Figure 5.10, and the algorithm of the minimization subprogram by scanning along the gradient direction of the function is shown in Figure 5.11.

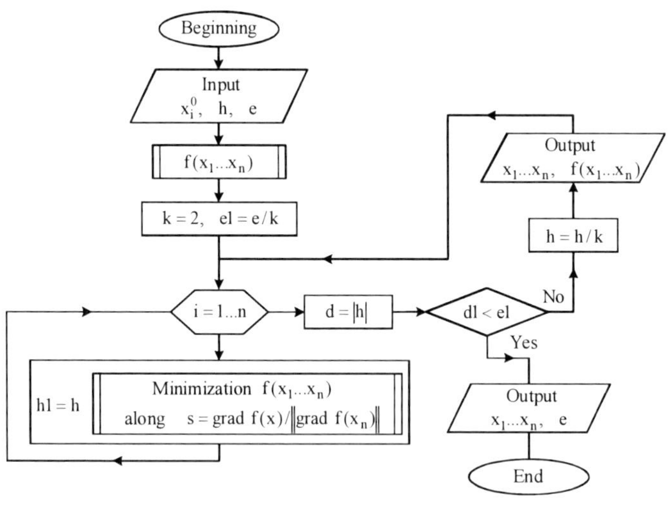

Figure 5.10. The algorithm for finding the minimum function by the method of rapid descent.

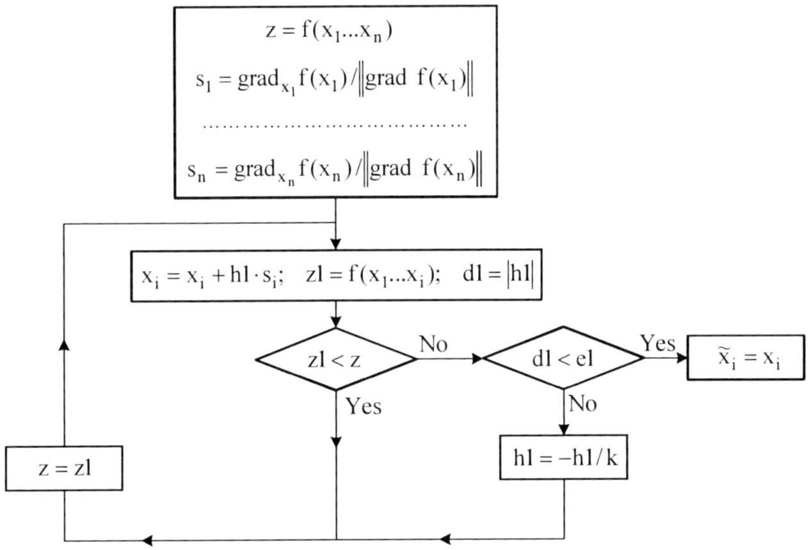

Figure 5.11. The algorithm of the minimization subprogram by scanning along the direction of the function gradient.

The termination of the computational procedure ends when the specified accuracy e is reached, which is replaced by almost equivalent condition $|h| < e$, especially after achieving high accuracy. Step h is a start change of the parameter in the minimization task, which decreases from one stage of calculations to another in k times. Minimization takes place with the decreasing step h_1 up to meeting condition $|h_1| < e_1$. Accuracy e_1 is taken slightly higher in comparison with e : $e_1 = e / k$.

However, as noted earlier, when IM is supplied from the frequency converter, the motor windings in most cases are connected to form a "star" without a neutral wire and two phases are involved in the work at the same time. Therefore, in the identification mode, the parameters of not one but two phases of the motor are determined. As at the adjustment of the systems of vector control it is required to have the information on electromagnetic parameters of the motor phases, in the identification mode it is necessary to carry out recalculations from the interphase to the phase parameters. In turn, when setting up a separate control system, in addition to the phase parameters of the windings, it is also necessary to determine the levels of asymmetry of the stator windings used in the control system to compensate for the variable component of electromagnetic torque or power consumption. Therefore, it is expedient to solve the system of equations with respect to the basic value of the parameter. The resistance of an arbitrary phase and the degrees of asymmetry in other two phases may represent such a parameter.

Thus, when the two phases operate simultaneously, the resistances of the stator windings are determined as follows [236]:

$$R_{AB} = R_A + R_B;\; R_{BC} = R_B + R_C;\; R_{AC} = R_A + R_C. \tag{5.17}$$

Assuming that $R_A = R$, $R_B = \varepsilon_{w1} R$, $R_C = \varepsilon_{w2} R$, where $\varepsilon_{w1}, \varepsilon_{w2}$ – the coefficients of asymmetry of the motor phases can be written:

$$\begin{cases} R_{AB} = R(1+\varepsilon_{w1}); \\ R_{BC} = R(\varepsilon_{w1}+\varepsilon_{w2}); \\ R_{AC} = R(1+\varepsilon_{w2}). \end{cases} \tag{5.18}$$

Solving this system of equations allows obtaining the following relations:

$$\begin{cases} R = \dfrac{R_{AB}+R_{AC}-R_{BC}}{2}; \\ \varepsilon_{w1} = \dfrac{R_{AB}-R_{AC}+R_{BC}}{R_{AB}+R_{AC}-R_{BC}}; \\ \varepsilon_{w2} = \dfrac{R_{AC}-R_{AB}+R_{BC}}{R_{AB}+R_{AC}-R_{BC}}. \end{cases} \tag{5.19}$$

Thus, when solving this system of equations, the resistances of the stator windings and the levels of asymmetry of the windings used to set up a variable-frequency electric drive with phase control are determined, according to which $\varepsilon_{\Psi a} = 1$, $\varepsilon_{\Psi b} = \varepsilon_{w1}$, $\varepsilon_{\Psi c} = \varepsilon_{w2}$. However, coefficients ε_{w1} and ε_{w2} may take values greater than 1, and there may be two variants when $\varepsilon_{w1} > \varepsilon_{w2}$ and when $\varepsilon_{w1} < \varepsilon_{w2}$:

- in the former case ($\varepsilon_{w1} > \varepsilon_{w2}$) – the calculation is carried out according to expressions: $R^* = \varepsilon_{w1} R$, $\varepsilon_{w1}^* = 1/\varepsilon_{w1}$, $\varepsilon_{w2}^* = \varepsilon_{w2}/\varepsilon_{w1}$, then during adjusting the variable-frequency electric drive the asymmetry coefficients by phases are determined as $\varepsilon_{\Psi a} = \varepsilon_{w1}^*$, $\varepsilon_{\Psi b} = 1$, $\varepsilon_{\Psi c} = \varepsilon_{w2}^*$.
- in the latter case ($\varepsilon_{w1} < \varepsilon_{w2}$) – the calculation is carried out according to expressions: $R^* = \varepsilon_{w2} R$, $\varepsilon_{w1}^* = 1/\varepsilon_{w2}$, $\varepsilon_{w2}^* = \varepsilon_{w1}/\varepsilon_{w2}$, then when adjusting the frequency-controlled electric drive, the phase asymmetry coefficients are determined as $\varepsilon_{\Psi a} = \varepsilon_{w1}^*$, $\varepsilon_{\Psi b} = \varepsilon_{w2}^*$, $\varepsilon_{\Psi c} = 1$.

Other parameters of IM equivalent circuit are recalculated according to the method presented in [237]. In accordance with it, the rotor resistance and the inductive resistance of the stator and rotor are determined by the expressions:

$$\left.\begin{aligned}
\chi_A &= (\chi_{AB} - \chi_{BC} + \chi_{AC})/2;\\
\chi_B &= (\chi_{AB} + \chi_{BC} - \chi_{AC})/2;\\
\chi_C &= (-\chi_{AB} + \chi_{BC} + \chi_{AC})/2,
\end{aligned}\right\} \tag{5.20}$$

where χ_A, χ_B, χ_C, χ_{AB}, χ_{BC}, χ_{AC} – respectively phase and interphase electromagnetic parameters of IM equivalent circuit.

To determine the inductance parameters of the magnetization circuit, taking into account that the interphase parameter X_μ is determined by the geometric sum of the phase parameters, we can write down:

$$\left.\begin{aligned}
X_{\mu AB} &= \sqrt{X_{\mu A}^2 + X_{\mu B}^2 - 2X_{\mu A} X_{\mu B}\cos(\varphi)};\\
X_{\mu BC} &= \sqrt{X_{\mu B}^2 + X_{\mu C}^2 - 2X_{\mu B} X_{\mu C}\cos(\varphi)};\\
X_{\mu AC} &= \sqrt{X_{\mu A}^2 + X_{\mu C}^2 - 2X_{\mu A} X_{\mu C}\cos(\varphi)},
\end{aligned}\right\} \tag{5.21}$$

where φ – the angle between the stator windings.

Solving this system of linear equations provides the determination of the parameters of the inductances of the magnetization circuit by IM phases. Thus, the presented recalculation procedure makes it possible to determine the phase parameters of IM equivalent circuit and the levels of asymmetry of the stator windings for further adjustment of the variable-frequency ED system.

The assessment of the adequacy of the obtained results is grounded on the method specified in [238, 239], which is based on the comparison of the output current signal and the one obtained from the output voltage and the calculated parameters. In case of signal mismatch, the system of equations is calculated again, but the values of the parameters obtained in the previous calculations are used as initial approximations.

5.3. The Systems for Measuring the Electrical Parameters of the Variable-Frequency Electric Drive

In the conditions of the stator windings power supply by the voltage generated by the method of pulse-width modulation, the problem of the substantiation of the parameters of measuring sensors and the module of analog-to-digital conversion (ADC) requires solution.

The substantiation of choosing the ADC module parameters is based on the decomposition of the voltage signal from the PWM of the stator windings, in the Fourier series with its subsequent restoration (by way of example, 8 kHz modulation frequency is chosen). The coefficient of determination can be used to find the number of harmonics for which the most accurate signal reproduction is observed. Based on the values of this factor for different drops from 0.1 to 0.9, the fullest reproduction of the signal occurs when using 11 harmonics of the voltage signal (Figure 5.12).

Fourier series expansion and construction of AFC of the initial function (Figure 5.13) also shows that significant voltage harmonics are in the range of 1 to 11. Based on Kotelnikov's theorem [198], which is a theoretical justification for the possibility of transmitting continuous signals by discrete values, we obtain $f_{ADC} = 2f_{edge}$. It follows from the obtained results that for the high-quality measurement the voltage of FC with PWM with a carrier frequency of 8 kHz the sampling frequency per ADC channel must be at least 176 kHz, which will cause the processing of unreasonably large amounts of data, even within one period of network frequency. The measurement of signals on six channels (voltage and currents of three phases of the stator) with the specified frequency makes the measurement system expensive and inefficient.

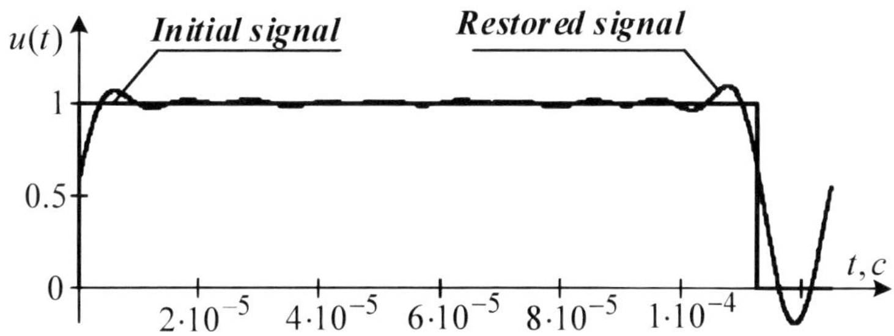

Figure 5.12. The initial and restored voltage signals generated by the PWM method.

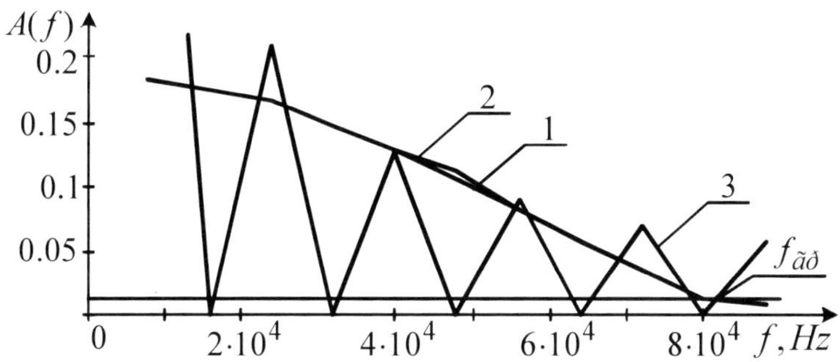

Figure 5.13. AFC of the voltage signal at different pulse filling coefficients: 1 – 0.1; 2 – 0.9; 3 – 0.5.

The problem of measuring the voltage of the stator windings of the variable-frequency ED can be solved by using the voltage signal of the DC link and the control signals of the power switches. However, this approach has a significant drawback, namely the emergence of significant errors due to failure to take into account the delay time of the keys and the so-called "dead time" – the delay time when switching transistors of one arm, generated by power transistor drivers.

In turn, a simple and economical way to increase the accuracy of measurements is to install low-frequency filters (LFF) in front of the measuring sensors with subsequent digital correction of the signals (Figure 5.14). The properties of this filter must provide the transmission of the low-frequency components of the input signal to the output with minimal distortion, while the high-frequency components, up to infinitely high frequencies, are to be delayed [198].

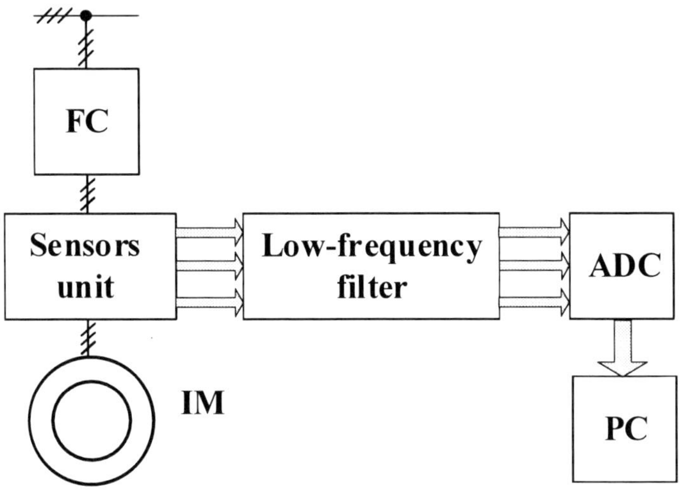

Figure 5.14. The functional diagram of the implementation of measurements in the supply of IM from FC.

The structure of the measuring channel is shown in Figure 5.15. The voltage sensors must be made with a variable coefficient to measure the current and voltage signals in a wide range.

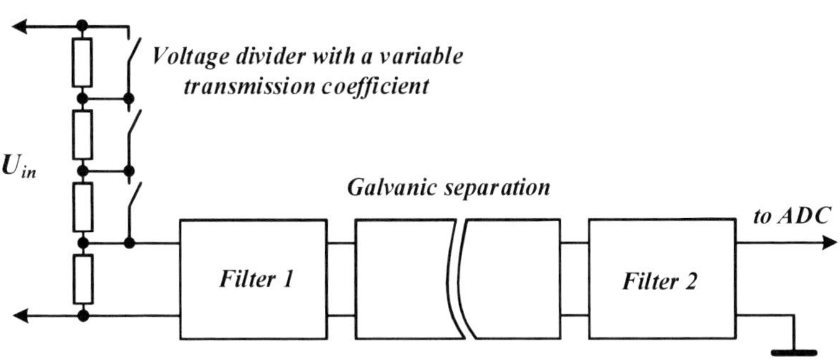

Figure 5.15. The block diagram of the measuring channel.

Based on the above, the voltage is measured as follows:

- filtering the signal after the voltage divider using a low-frequency filter;
- galvanic separation of power and measuring circuits by means of the chip of the amplifier with galvanic separation;
- the second stage of filtration based on the passive RC circuit;
- signal output to the ADC;
- the expansion of the Fourier series and software signal correction based on the known expressions for AFC and PFC filters. Moreover, not the whole signal is restored, but only the first, and if necessary, the third and fifth harmonics.

Filter design is a search for a compromise between high performance and the complexity of its implementation.Taking into account that in this case the requirements for the quality of filtering allow one to do with the simplest filters of the first or second order, the design of the filter is reduced to choosing a scheme from the existing options with the simplest and most rational configuration for specific frequencies [240, 241]. As the designed filter is focused on measuring signals with the fundamental harmonic at a frequency of 50 Hz and below, a fourth-order filter with a cutoff frequency of 300 Hz can be used for the specified method of measurement, based on the requirement of minimum amplitude and phase distortion at low frequencies. Its creation requires two second-order links whose circuits are shown in Figure 5.16.

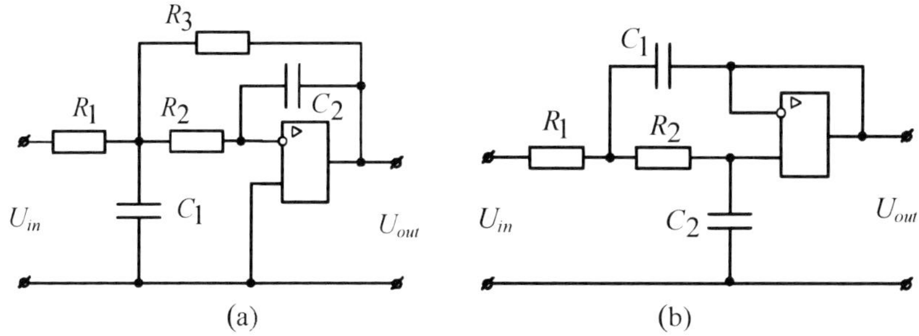

Figure 5.16. The circuits of low-frequency filters of the second order: a) – the first, b) – the second stage of filtration.

The specified filter inserts amplitude and phase distortions, the presence of which in the experimentally measured signals introduces significant errors in further calculations [238, 242]. These distortions can be assessed and corrected on the basis of the analysis of AFC and PFC of the selected units (Figure 5.17) obtained from the transmission characteristics of the filter [243, 244]. Table 5.4 contains the values of the amplitude and phase distortions of the filters.

Table 5.4. The values of the amplitude and phase distortions of the filters

Frequency. Hz	AFC value		PFC value	
	First link	Second link	First link	Second link
50	0.99975	0.99978	-1.4198	-1.6615
150	0.99777	0.99806	-4.2543	-4.9813
250	0.99385	0.99462	-7.0741	-8.2921
350	0.98805	0.98951	-9.8697	-11.588

The transmission functions of the two links are determined as follows:

- the first link:

$$W_1(s) = \frac{1}{C_1C_2R_1R_2s^2 - [(R_1 + R_2)C_2 + R_1R_2C_2/R_3]s - R_1/R_3};$$

- the second link:

$$W_2(s) = \frac{1}{C_1C_2R_1R_2s^2 - (R_1 + R_2)C_2s - 1}.$$

The correction of the measured voltage signals is performed (Figure 5.18) based on the given values of attenuation of harmonic amplitudes and phase shift on harmonics introduced by filters (Table 5.4).

Table 5.5 contains the corrected values of the amplitudes and phases of the current and voltage signals.

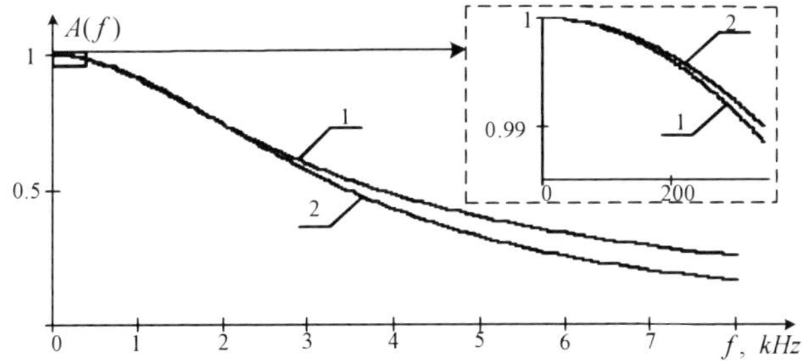

(a)

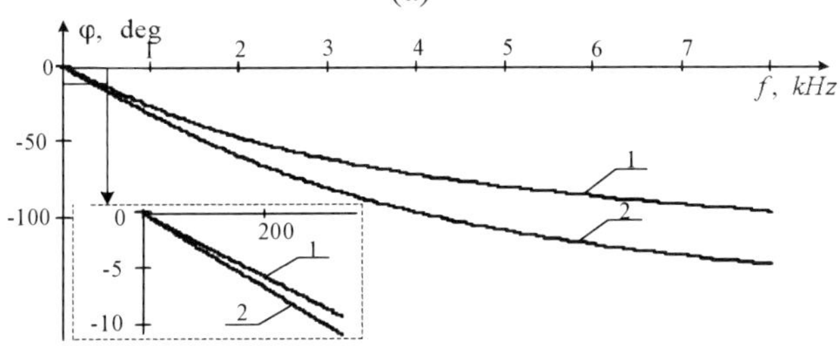

(b)

Figure 5.17. AFC (a) and PFC (b) of the second-order filters: 1 – the first link; 2 – the second link.

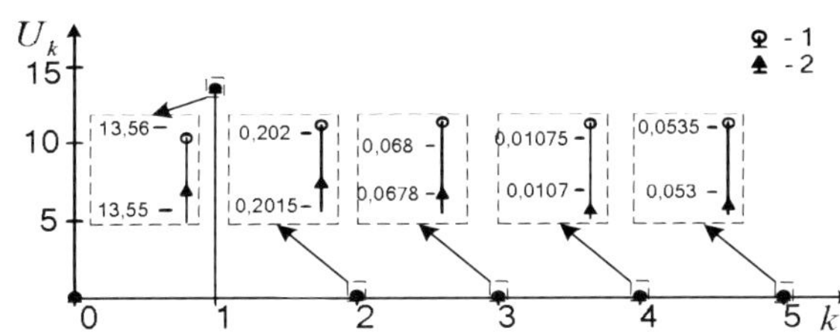

(a)

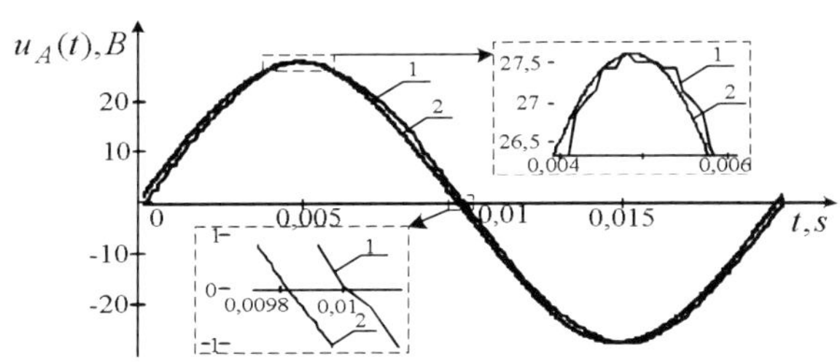

(b)

Figure 5.18. The spectral composition (a) and correction of the voltage signal (b) in phase A before (1) and after (2) correction.

Table 5.5. The corrected values of the amplitudes and phases of the frequency components of the voltage signal

	Frequency, Hz					
	Constant component	**50**	**100**	**150**	**200**	**250**
Amplitude	0.038	13.552	0.2017	0.0679	0.0107	0.0529
Phase	0	-86.79	-72.30	96.636	-60.58	58.768

The experimental signals of IM stator windings voltage, when powered by a frequency converter, obtained using the proposed measurement system, are shown in Figure 5.19.

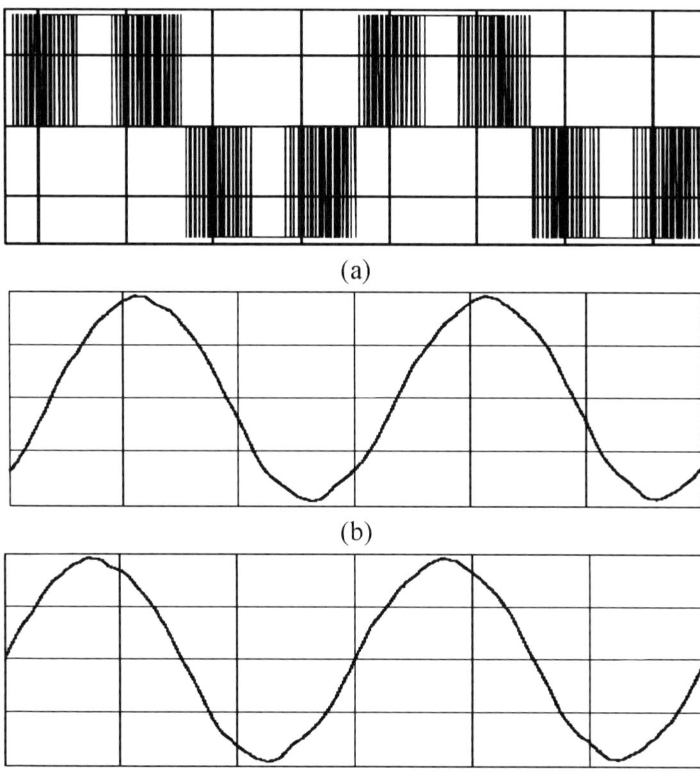

(a)

(b)

(c)

Figure 5.19. The experimental signals of IM stator voltage at the output of the frequency converter (a), from the output of the low-frequency filter (b) and after the digital correction (c).

Thus, the substantiated parameters of the measuring system of the variable-frequency ED, in which the measurement of the voltage of IM stator phases is based on the use of low-frequency filters with subsequent digital correction of the signal, make it possible to increase the measurement accuracy.

5.4. The Experimental Verification of the System of the Assessment of the Parameters of the Induction Motor in the Pre-Start Period

The experimental verification of the proposed method for determining the parameters of IM equivalent circuit, using low-frequency supply voltage, can be provided on the basis of the proposed structures of test equipment [245], the functional diagrams of which are given in Figure 5.20.

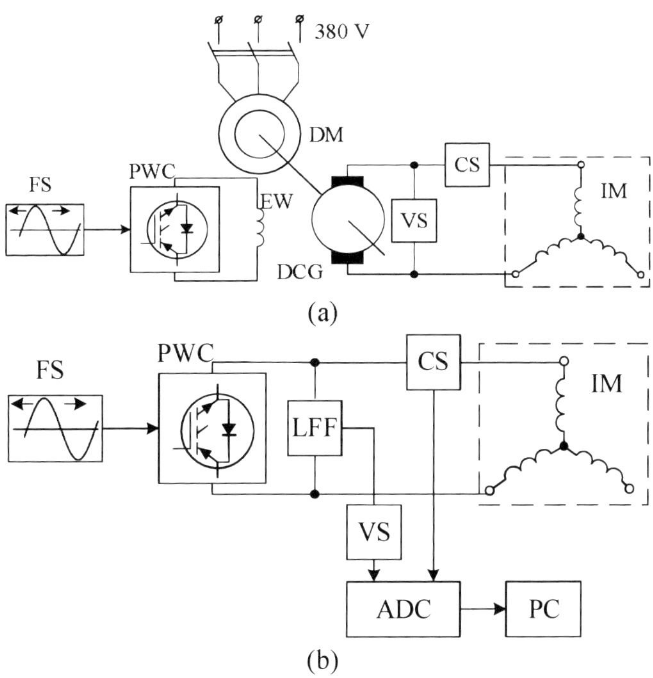

Figure 5.20. The functional diagrams of the structure of the research equipment: FS – frequency setting; PWC – power converter; EW – excitation winding; DM – driving motor; DCG – direct current generator; LFF – low frequency filter; VS – voltage sensor; CS – current sensor.

In the first case a direct current generator (DCG), whose excitation winding is equipped with a power frequency converter (PWVC), is used (Figure 5.20, a). In the other case, the test motor windings are fed directly from the voltage inverter (Figure 5.20, b), and to increase the accuracy of measuring current and voltage signals, a low-pass filter is installed before the measuring sensors [246]. In this case, the tests are performed on the basis of a short-circuit experiment, the essence of which is as follows: the PC feeds low-frequency supply voltage to DCG excitation winding, and the armature is connected to the phase windings and the current and voltage signals are recorded [247]. The functional diagram of the experimental equipment using a power converter with a formed sinusoidal test voltage is shown in Figure 5.21.

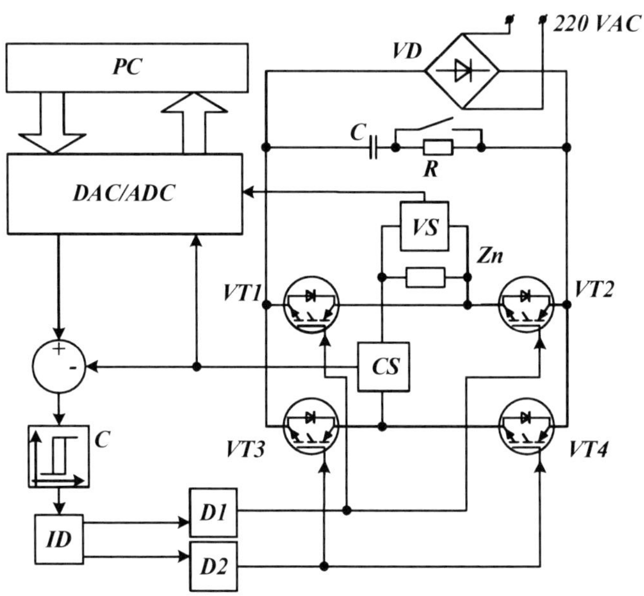

Figure 5.21. The functional diagram of a low-frequency supply source: C – comparator; ID – impulse distributor; D1, D2 – the first and the second driver of the power keys control; VT1-VT4 – power transistor keys; VS – voltage sensor; CS – current sensor.

This converter is based on the principle of "current corridor" [7], in which the load current fluctuates around a given control signal (Figure 5.22). The power keys of the converter are switched by means of comparator *C*. In turn, the state of the signal at the output of the comparator depends on the sign of the deviation of the actual load current signal from the reference signal. The average value of the load current corresponds to the specified value with error $\pm\Delta$, determined by the value of the hysteresis comparator C. The power keys of the converter are switched by means of comparator C. In turn, the state of the signal at the output of the comparator depends on the sign of the deviation of the actual load current signal from the reference signal. The average value of the load current corresponds to the specified value with error.

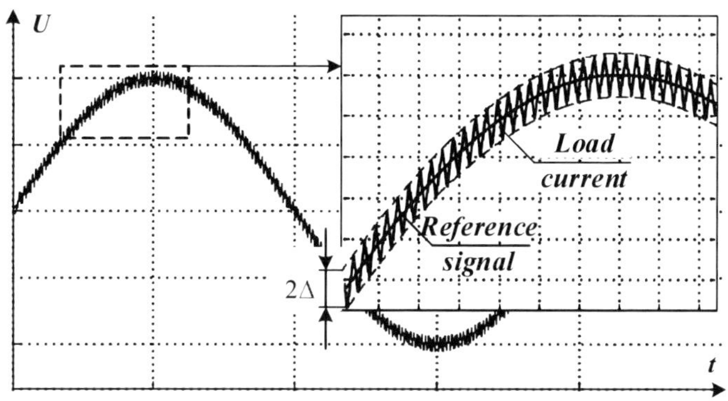

Figure 5.22. The development of the reference signal by the converter, with "current corridor" control.

In the shown system, the reference signal is added to the current signal flowing through the load by means of an adder (Figure 5.23).

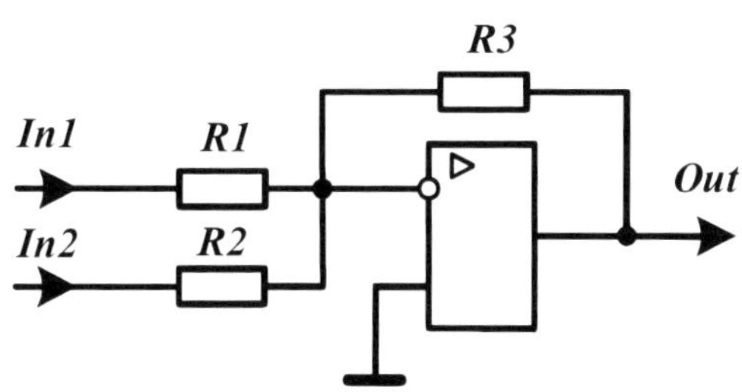

Figure 5.23. The adder on the operational amplifier.

Since the output signal of the adder is an analog signal, it is advisable to use integrated comparators to convert it into a discrete form (Figure 5.24). An input signal is applied to one of the inputs of the comparator, and a reference voltage is applied to the other. When the input signal exceeds the reference voltage, the comparator is switched. To increase the noise immunity of the circuit, positive feedback is introduced to it, and hysteresis appears in the initial characteristic of the comparator.

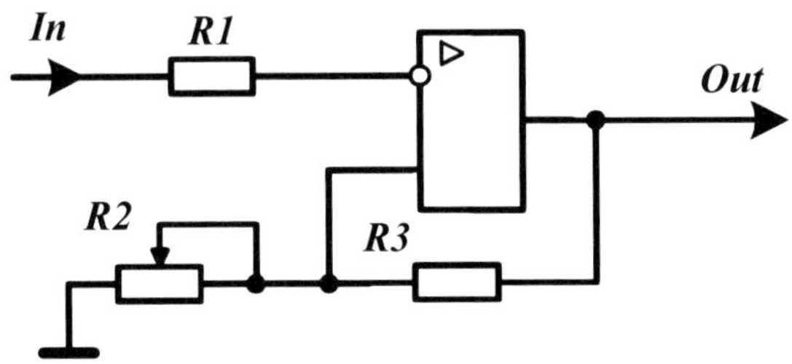

Figure 5.24. Integral comparator with hysteresis.

Output pulses from the comparator are fed to the input of the pulse distributor (Figure 5.25), which is performed on the basis of logic elements. Elements VD_1, R_1, C_1 and VD_2, R_2, C_2 form a protective pause in the output control signals of the half-bridge transistors.

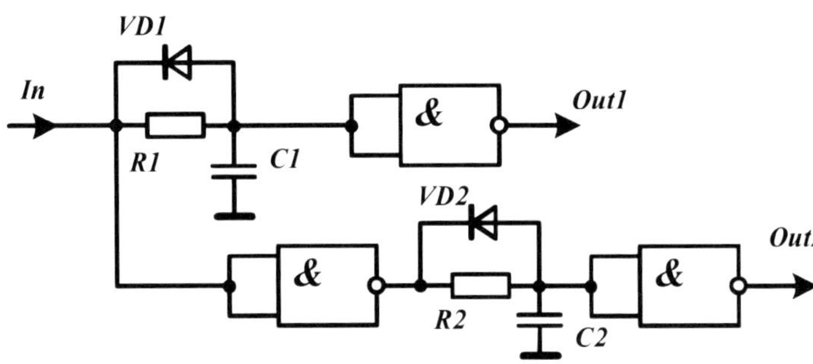

Figure 5.25. Pulse distributer.

An experimental verification of the proposed method for determining the electromagnetic parameters of IM replacement circuit was performed for 4AHB2P100L motor ($P_n = 4$ kW, $n_n = 1420$ rev/min, $cos\,\varphi = 0.84$, $\eta = 84\%$, $I_n = 15$ A, $R_A = 1.98$ Ohm, $R_B = 1.96$ Ohm. The assessment of the deviation of the shape of the experimental signals of current and voltage from the sine wave is performed by the non-sine wave coefficients [188]:

$$k_u = \frac{\sqrt{\sum_{i=2}^{n} (U_i)^2}}{U_1} \cdot 100\,\%;$$

$$k_i = \frac{\sqrt{\sum_{i=2}^{n} (I_i)^2}}{I_1} \cdot 100\,\%.$$

The parameters of the experimental signals with connected phase A, phase B and phase AB are given in Table 5.6.

Table 5.6. The results of the calculation of the parameters of the current and voltage signals by the phases of the researched IM

U_3, **V**	***f*, Hz**	U_1	I_1	$\mathrm{Re}\left(\frac{U_1}{I_1}\right)$	$\mathrm{Im}\left(\frac{U_1}{I_1}\right)$	k_u	k_i
Phase A							
1	2	1.66-8.31i	-0.09-4.34i	1.908	0.422	2.12	1.87
0.5	1	1.29-8.56i	-0.25-4.72i	1.794	0.369	2.47	2.27
0.5	0.75	-1.05+8.58i	0.31+5.34i	1.59	0.289	3.57	2.19
Phase B							
1	2	2.84-13.75i	-0.08-7.14i	1.923	0.419	5.87	4.95
0.5	1	1.87-8.5i	0.05-4.76i	1.789	0.373	2.18	2.39
0.5	0.75	1.05-8.64i	-0.03-5.63i	1.534	0.195	3.71	1.82
Phase AB							
2	2	2.42-18.74i	-0.29-4.32i	4.28	0.854	7.98	7.91
1	1	3.25-14.65i	-0.07-3.8i	3.838	0.93	8.31	9.83
1	0.75	1.95-14.86i	-0.35-4.57i	3.198	0.674	10.1	6.58

The experimentally measured current, voltage signals and their spectral composition with connected phase A of the researched motor at a frequency of 2 Hz are shown in Figure 5.26, and with connected phase B at a frequency of 1 Hz they are shown in Figure 5.27.

The resulting current and voltage signals are expanded into a Fourier series. The values of the first harmonics U_1 and I_1 (Table 5.6) are used to calculate the value of the complex impedance, in which the real part

$\text{Re}(U_1/I_1)$ is a total resistance of the windings and the imaginary part $\text{Im}(U_1/I_1)$ – a total inductive reactance of the windings of the researched motor. The dependences of the change of the calculated total values of the researched motor resistance and inductive reactance on the frequency, with connected phases A and B are shown in Figure 5.28, a, and with connected two phases (AB) are shown in Figure 5.28, b.

On the basis of experimentally obtained current and voltage signals, at low frequencies, the parameters of IM equivalent circuit are calculated for each phase separately (connection circuit – a star with zero) and for two phases (connection circuit – a star without zero). The calculation results are presented in Table 5.7.

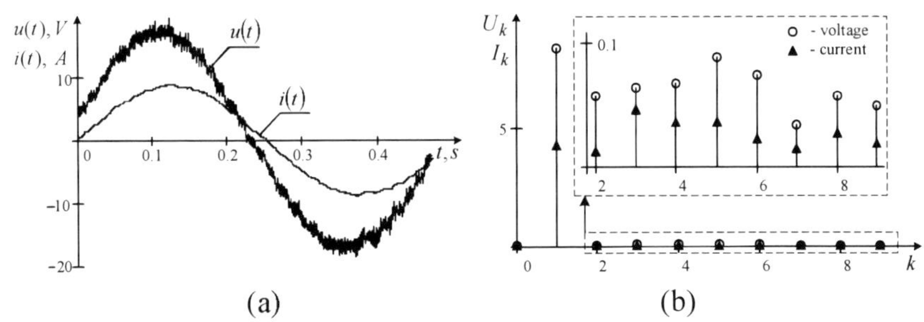

Figure 5.26. The experimental curves of phase A (a) voltage and current and their spectral composition (b).

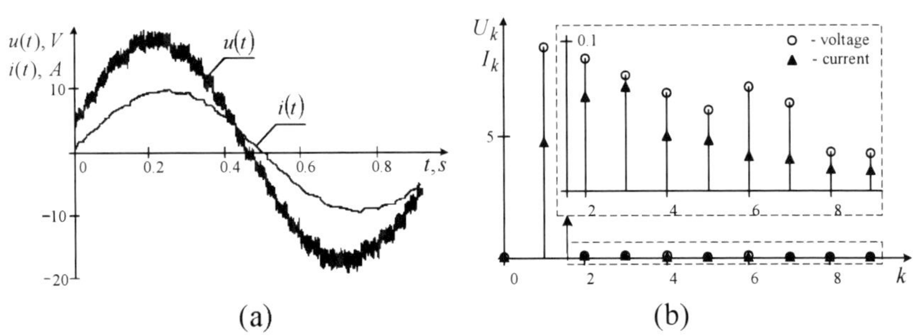

Figure 5.27. The experimental curves of phase B (a) voltage and current and their spectral composition (b).

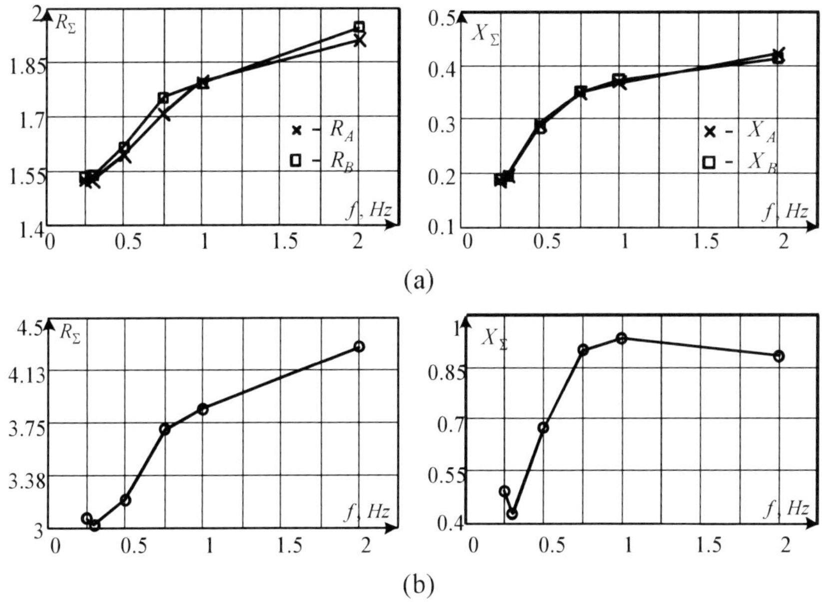

Figure 5.28. The dependences of the total resistances and inductive reactance on the supply voltage frequency for phases A and B (a) and AB (b).

Table 5.7. The calculated parameters of the researched motor phases

Parameter	Phases		
	A	B	AB
R_r , Ohm	1.1449	1.1799	2.4924
L_s , Ohm	0.0065	0.0065	0.0131
L_r , Ohm	0.0068	0.0068	0.0137
L_μ , Ohm	0.1135	0.1275	0.2639

The results of the experimental verification of the proposed identification method revealed that the errors in determining IM electromagnetic parameters do not exceed 2% for resistances, 1.5% for scattering inductances, 4% for the main mutual inductance.

The accuracy of determining the electromagnetic parameters is assessed by comparison of the experimental signals of currents and the signals calculated with use of the obtained parameters of the equivalent circuit. The experimental voltage curves were used to obtain the calculated currents. Thus, the comparison of current signals of the researched motor phase A at frequencies of 50 and 2 Hz is shown in Figure 5.29, phase B at frequencies of 50 and 1 Hz is shown in Figure 5.30.

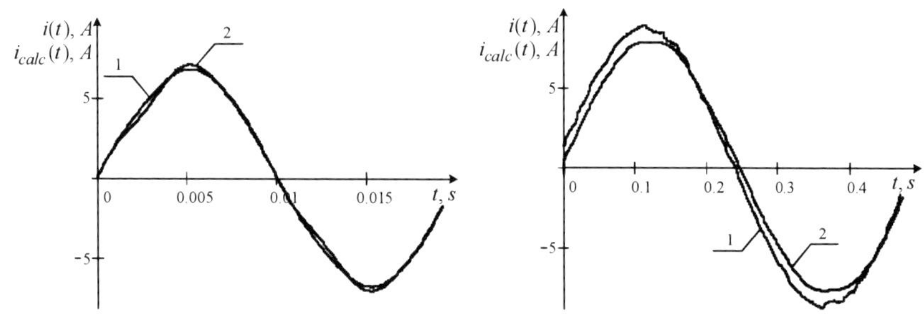

(a) (b)

Figure 5.29. The comparison of the output (1) and calculated (2) signals of phase A current at frequencies of 50 Hz (a) and 2 Hz (b).

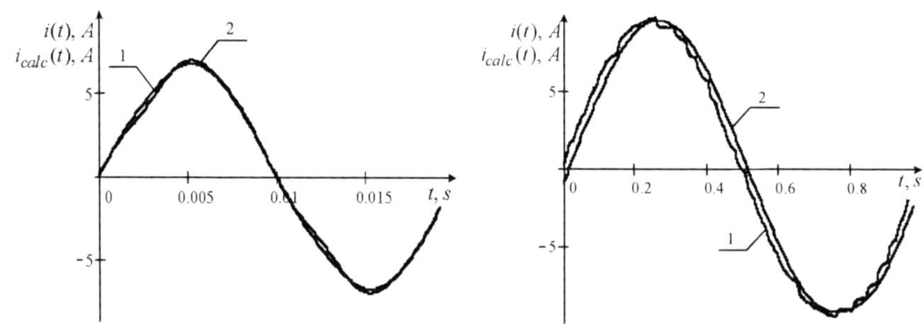

(a) (b)

Figure 5.30. The comparison of the output (1) and calculated (2) signals of phase B current at frequencies of 50 Hz (a) and 2 Hz (b).

The experimental research revealed that the solution to the system of nonlinear equations by the steepest descent method allows determining IM parameters with high accuracy with the application of reasonable initial approximations of electromagnetic parameters, which significantly narrow the range of the search for unknown parameters, and the use of the signal amplitude and phase as the criteria for the coincidence of the experimental and the calculated current signals.

5.5. The Influence of the Signal Amplitude and Phase Errors on the Determination of Induction Motor Electromagnetic Parameters

The amplitude and phase of the signal can be used as the criteria for the coincidence of the experimental current signal with the calculated one, obtained on the basis of certain parameters of IM equivalent circuit and experimental voltage signals [198] (Figure 5.31).

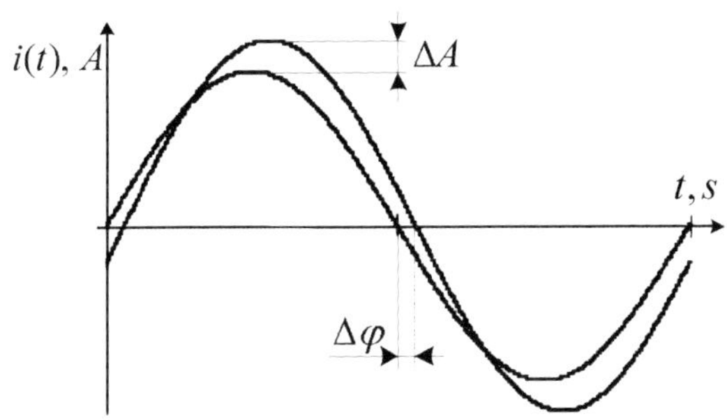

Figure 5.31. The errors of the calculated reproduction of the current sinusoidal signals.

These criteria can be used to mathematically describe the error of identification in the form of objective functions [248], which makes it possible:

- first, to find out which of the identified parameters affect the accuracy of the current signal reproduction;
- second, to identify how the errors of these criteria in the experimentally measured signals affect the accuracy of determining the parameters of IM equivalent circuit.

Thus, the construction of objective functions is reduced to two cases – the definition of the quality criteria by amplitude and phase.

In the first case, the objective function is determined by the root-mean-square deviation of the current amplitude, which is defined as the sum of the squares of the ratio of the difference between the experimental and calculated

I_{en} and calculated I_{pn} value of the current signal to the experimental value obtained at the corresponding reduced supply frequencies:

$$J_A = \sum_{n}^{N} \left(\frac{I_{en} - I_{pn}}{I_{en}} \right)^2 , \tag{5.25}$$

where J_A – the objective function by amplitude quality criterion; N – the number of low-frequency power experiments; I_{en} – the experimental values of current amplitude at different frequencies; I_{pn} – the calculated value of the current amplitude at the corresponding frequencies of the power supply network. The calculated value of the current at low frequencies can be defined as:

$$\dot{I}_n = \dot{U}_n / \dot{Z}_n ,$$

where $\dot{U}_n$ – the first harmonic of the experimental voltage signal at a correspondingly reduced power frequency (in this case the frequencies are 2. 1 and 0.75 Hz); $\dot{Z}_n$ – IM complex impedance [249] at reduced frequencies:

$$\dot{Z}_n = R_s + \frac{R_r (L_\mu \omega_n)^2}{R_r^2 + (L_{r\sigma}\omega_n + L_\mu \omega_n)^2} + \\ + j \left[L_{s\sigma}\omega_n + \frac{R_r^2 L_\mu \omega_n + L_\mu \omega_n (L_{r\sigma}\omega_n)^2 + (L_\mu \omega_n)^2 L_{r\sigma}\omega_n}{R_r^2 + (L_{r\sigma}\omega_n + L_\mu \omega_n)^2} \right],$$

where ω_n – the angular frequency at the corresponding voltage frequencies.

The influence of the errors in determining the electromagnetic parameters of IM equivalent circuit, with their deviation from the rated values by 50-150%, on the accuracy of the calculated reproduction of the current signal by amplitude is shown in Figure 5.32.

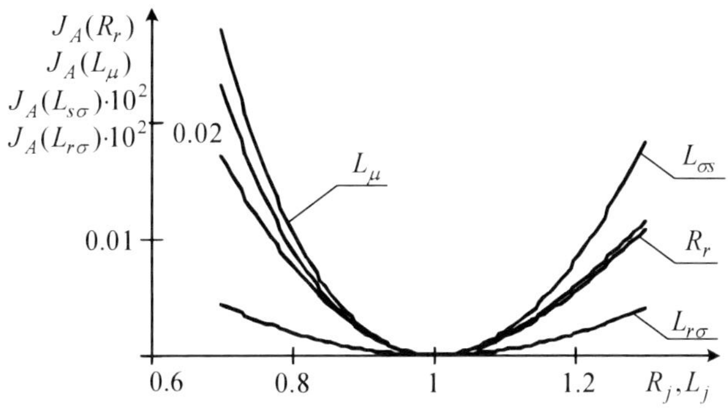

Figure 5.32. The change of the RMS quality criterion by the amplitude.

Thus, the accuracy of reproducing the amplitude of the current signal is more affected by the accuracy of determining the values of the rotor resistance and inductance of the magnetization circuit, and to a lesser extent – by the values of the scattering inductances of the stator and rotor.

The formation of the objective function for the other case is based on the root-mean-square deviation of the quality criterion by phase, which is defined as the sum of the squares of the difference between the experimental φ_{en} and calculated φ_{pn} values of the phases of the signals obtained at the corresponding reduced supply frequencies:

$$J_\varphi = \sum_n^N \left(\frac{\varphi_{en} - \varphi_{pn}}{\pi} \right)^2, \tag{5.26}$$

where J_φ – the objective function by the phase quality criterion; φ_{pn} – the calculated value of the phase of the researched signals at different frequencies:

$$\varphi_{pn} = arctg \frac{\left(L_{s\sigma}\omega_n + \dfrac{R_r^2 L_\mu \omega_n + L_\mu \omega_n (L_{r\sigma}\omega_n)^2}{R_r^2 + (L_{r\sigma}\omega_n + L_\mu \omega_n)^2} + \dfrac{(L_\mu \omega_n)^2 L_{r\sigma}\omega_n}{R_r^2 + (L_{r\sigma}\omega_n + L_\mu \omega_n)^2} \right)}{R_s + \dfrac{R_r (L_\mu \omega_n)^2}{R_r^2 + (L_{r\sigma}\omega_n + L_\mu \omega_n)^2}};$$

φ_{en} – the experimental value of the phase of the researched signals:

$$\varphi_{en} = arctg\left(\mathrm{Im}\left(\dot{U}_n\right)/\mathrm{Re}\left(\dot{U}_n\right)\right) - arctg\left(\mathrm{Im}\left(\dot{I}_n\right)/\mathrm{Re}\left(\dot{I}_n\right)\right),$$

where $\mathrm{Re}\left(\dot{U}_n\right)$, $\mathrm{Im}\left(\dot{U}_n\right)$, $\mathrm{Re}\left(\dot{I}_n\right)$, $\mathrm{Im}\left(\dot{I}_n\right)$ – the real and the imaginary parts of the voltage and current experimental components.

The influence of the errors in determining the electromagnetic parameters of IM equivalent circuit on the accuracy of the calculated reproduction of the current signal by phase is shown in Figure 5.33.

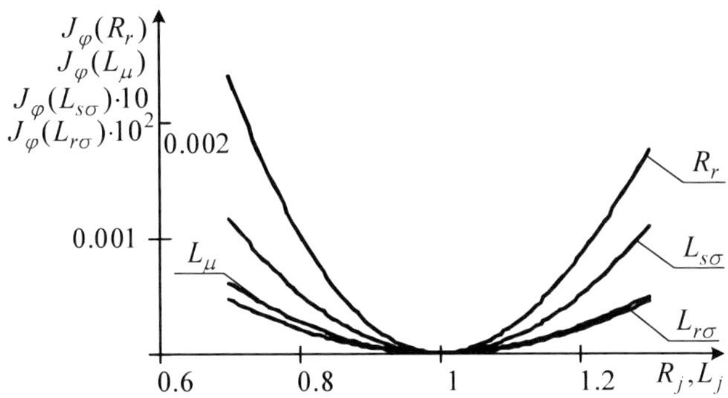

Figure 5.33. The change of the RMS quality criterion by phase.

Therefore, the accuracy of the reproduction of the signal phase, as well as its amplitude, is more affected by the accuracy of determining the values of the rotor resistance and inductance of the magnetization circuit, and less by the values of the stator and rotor scattering inductances.

When researching IM operation modes at the rated and higher frequencies of the supply voltage, the influence of the magnetization circuit is sometimes neglected, then the equation of the calculated value of the modulus of complex impedance takes the form:

$$\dot{Z}_n = \sqrt{\left[R_s + R_r\right]^2 + \left[L_{s\sigma}\omega_n + L_{r\sigma}\omega_n\right]^2}\,.$$

In this case, the change in the RMS amplitude quality criterion with the deviation of the electromagnetic parameters of IM equivalent circuit from the rated values in the range of 50–150% for voltage signals at low frequencies and mains frequency is presented in Figure 5.34.

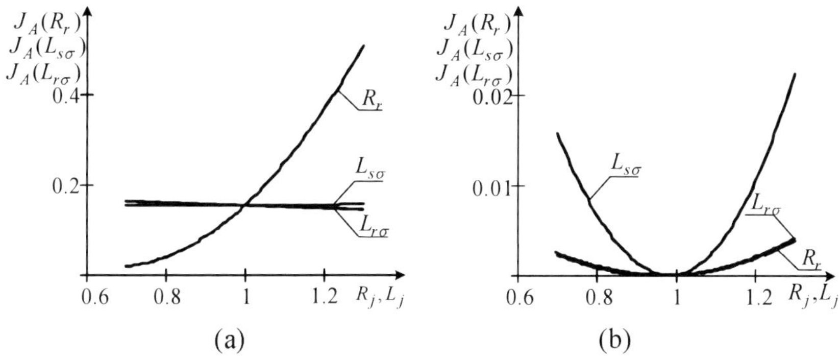

Figure 5.34. The change of RMS amplitude quality criterion at low frequencies (a) and mains frequency (b).

Under the accepted assumption, the equation of the calculated value of the phase takes the form:

$$\varphi_{pn} = arctg\left(\frac{L_{s\sigma}\omega_n + L_{r\sigma}\omega_n}{R_s + R_r}\right).$$

The change in the RMS quality criterion by the phase with the deviation of IM electromagnetic parameters from the rated values in the range of 50–150% for voltage signals at low frequencies and mains frequency is presented Figure 5.35.

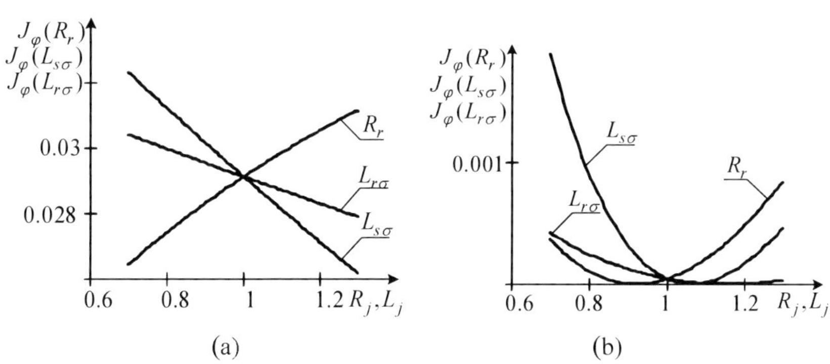

Figure 5.35. The change in the RMS quality criterion by the phase at low frequencies (a) and mains frequency (b).

The presented results indicate that in the short-circuit mode the inductance of the magnetization circuit cannot be neglected, because it increases the error of identification of electromagnetic parameters of the induction motor equivalent circuits.

The influence of the static error in the experimentally measured signal on the amplitude and phase quality criteria can be determined by adding the deviation in the amplitude and phase to the current signal:

$$I_{vn} = \sqrt{\mathrm{Re}(I_n)^2 + \mathrm{Im}(I_n)^2} \cdot \Delta A \cdot e^{i\left(arctg\left(\frac{\mathrm{Im}(I_n)}{\mathrm{Re}(I_n)}\right)\cdot\Delta\varphi\right)}, \tag{5.27}$$

where ΔA – the static error in the signal amplitude; $\Delta\varphi$ – the static error in the signal phase.

Thus, the influence of static measurement errors on the quality proposed criteria for low frequency and mains voltage signals is shown in Figure 5.36.

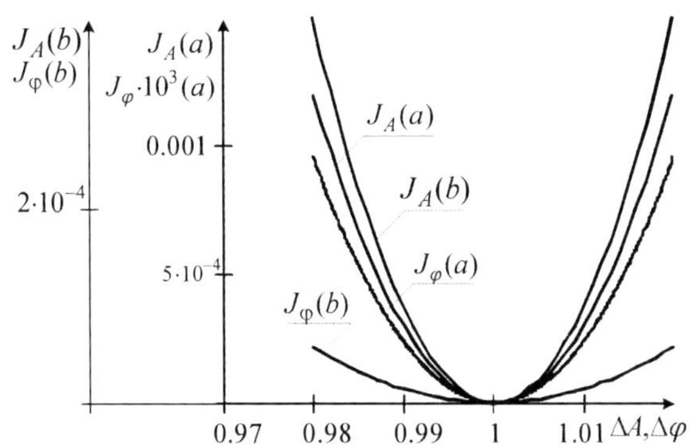

Figure 5.36. The influence of the static measurement error on the RMS quality criterion by the amplitude and phase at low frequencies (a) and mains frequency (b).

Thus, the presence of the static error in the amplitude of the signal has a greater effect on the accuracy of signal reproduction than the presence of the static error in the phase of the signal both at low frequency values and at a frequency of 50 Hz.

Table 5.8 contains the amplitude and phase deviations of current signals calculated by the output voltage and determined electromagnetic parameters of the motor (Table 5.7) by phases A, B, AB, from the experimentally measured signals.

Table 5.8. The deviations of the calculated current signals by amplitude (A) and phase (φ)

Frequency, Hz	Criterion		Errors by phases		
			A	B	AB
50	A, %		1.185	0.765	0.389
	φ,	deg	1.7	1.6	1.3
		%	0.94	0.89	0.72
2	A, %		10.257	11.743	3.129
	φ,	deg	5.2	5.2	6.7
		%	2.9	2.9	3.7
1	A, %		1.624	0.56	5.42
	φ,	deg	6.1	6.8	5.5
		%	3.4	3.8	3.1

Conclusion

1. It is shown that the methods for determining the parameters of the induction motor equivalent circuit in the pre-start period as part of a variable-frequency electric drive are implemented using different test effects on the motor stator windings and based on assumptions about the symmetry of electrical and magnetic systems. This assumption is not always correct when using induction motors that have been in operation for a long time and have been repeatedly repaired.
2. A method for the identification of the interphase parameters of the induction motor equivalent circuit in the pre-start-up period at different-frequency sinusoidal supply of the stator windings with low-frequency voltage is offered. This method allows improving the accuracy of determining the inductance of the magnetization circuit due to increasing the magnetizing current almost to the level of the rotor current. The advantages of the proposed method include the simplification of the mathematical apparatus of identification and sufficient accuracy in determining the electromagnetic parameters through the use of iterative algorithms based on the comparison of experimental and calculated current signals.
3. A method for calculating the phase-to-phase parameters of induction motors is proposed. It enables determining the phase parameters of the equivalent circuit and asymmetry levels of the motor stator windings by solving a system of linear algebraic equations relative to

the basic parameter for further adjustment of separate vector control system.

4. The choice of initial approximations for the method of identification of electromagnetic parameters of induction motor equivalent circuit is substantiated. They are found by the analysis of the first harmonics of current and voltage in short-circuit modes to determine the parameters of stator and rotor windings, and the idle speed to determine the inductance of the magnetization circuit, which makes it possible to identify the exact values of the parameters.
5. The structure and parameters of the measuring system as part of the variable-frequency electric drive are substantiated. They make it possible to improve the accuracy of measurements due to the use of filters of low frequencies and the subsequent digital correction of the signals.
6. The developed engineering solution for the construction of a controlled power supply allows working out a wide range of reference signals to research the methods for identifying electromagnetic parameters.
7. The research on the changes in the criteria for the minimum discrepancy between experimental and calculated signals revealed the effect of measurement errors and errors in determining the parameters on the assessed reproduction of the experimental signal.
8. The analysis of the accuracy of parameter determination showed that in the short-circuit mode the inductance of the magnetization circuit cannot be neglected, as this results in an increase in the error in restoring both phase and signal amplitude while reducing the frequency of the supply voltage.
9. The results of experimental research showed the efficiency of the developed methods for determining the initial approximations and direct electromagnetic parameters of the induction motor equivalent circuit, the errors of which do not exceed 2% for resistance, 1.5% for scattering inductances, 4% for magnetization circuit inductances. The application of the developed methods in industrial specimens of variable-frequency electric drives with the proposed measuring system makes it possible to solve the problems of identification of electromagnetic parameters of the induction motor equivalent circuit in the pre-starting period for each phase separately.

Chapter 6

The Development of the Methods for the Indirect Determination of the Energy Characteristics of IM Operation and the Improvement of the Economic Efficiency of the Induction Motor Stock Operation

6.1. The Quantitative Assessment of the Error of the Calculation of the Active Instantaneous Power of Three-Phase Systems in the Serial Poll of ADC of Voltage and Current Channels

One of the issues in solving the problem of assessing VFED efficiency consists in the reliable determination of the indicators of the mode of its operation in general and IM in particular. The solution to this problem is impossible without accurate definition of the set of parameters that characterize the system. The active instantaneous three-phase power is one of the most important parameters. The existing theories of instantaneous power described above, which are used in compensation systems, require clear and reliable information on the components of the power signal for the correct calculation of the reference signals for active compensators.These signals are calculated based on real-time instantaneous current and voltage signals obtained with the help of measuring complexes (MC).

A mutually exclusive problem occurs during MC development: the measuring complex, on the one hand, must provide sufficient accuracy of measurement parameters, and on the other hand, its cost must be reasonable. In solving this problem, there are strict requirements for the accuracy of MC, including the analog-to-digital converter ADC as its main component. Signal conversion leads to certain errors in the parameters that were calculated based on the signals obtained from the ADC. Moreover, these errors are additive in nature, and their value in each case depends on the type of the used converter. As a result, there are a number of subtasks related to the need to develop methods for the improvement of MC accuracy, both at the stage of measurements and at the stage of data processing and calculation of the parameters [152, 250].

Despite the widespread application of ADCs with a serial poll of channels their use is characterized by a number of problems.Thus, when researching the operation of three-phase systems, the main characteristic parameters are calculated on the basis of instantaneous signals of currents and voltages of the stator phases of the motor. So, when a serial ADC operates, the polling information from analog channels appears with a certain time delay Δt between the polls of each channel. That is, when recording sequentially the signals of the motor phases, the beginning of the reference period of the current phase C is shifted by a period of time $5 \cdot \Delta t$ relative to the initial moment of phase A voltage. This case is called "the sequential registration of phases," and the case when voltage signals $(u_A(t), u_B(t), u_C(t))$ and currents $(i_A(t), i_B(t), i_C(t))$ are read – "the sequential registration of signals."

To analyze the effect of this time delay, the modeling of the calculation of the instantaneous active power was performed for a three-phase system with a star-zero connection:

$$p(t) = u_A(t)i_A(t) + u_B(t)i_B(t) + u_C(t)i_C(t) \tag{6.1}$$

and without zero:

$$p(t) = \frac{1}{3}\begin{bmatrix} u_{AB}(t)(i_A(t) - i_B(t)) + \\ + u_{BC}(t)(i_B(t) - i_C(t)) + u_{CA}(t)(i_C(t) - i_A(t)) \end{bmatrix} \tag{6.2}$$

Channel polling was performed for both of these variants of the sequence according to (6.1) and (6.2) during the research of power determination. To do this, delay Δt determined by the discretization frequency of the ADC is artificially introduced into the model in the ideal signals of currents and voltages of the stator phases. The calculations were performed for the angles between current and voltage $\varphi = 30^{\circ}$, which approximately corresponds to the rated mode of operation of a medium power IM, and $\varphi = 80^{\circ}$ corresponding to IM idle mode. The following values of ADC discretization frequencies on 6 measured channels were selected for modeling: 60, 100, 120,

180 kHz. The choice of such conditions for modeling is justified by the fact that most ADCs operate in this frequency range.

As the results of the modeling show, such a time delay adds a significant variable component to the calculated instantaneous active power signal (Figure 6.1). The harmonic analysis of the calculated curve showed that the second harmonic appears in the power spectrum, and its amplitude depends not only on the sampling frequency of the ADC, but also on the load power coefficient (Figure 6.2). This harmonic usually indicates the asymmetry of the parameters of the electric machine or power supply, but in this case it is imaginary, i.e., it appears only as a calculation error.

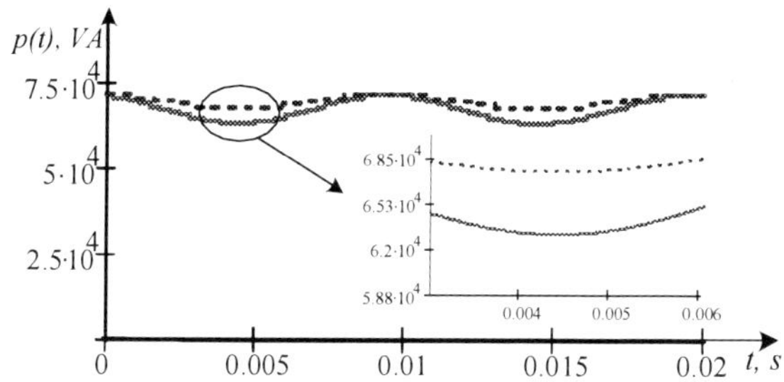

Figure 6.1. IM instantaneous three-phase active power with different settings of ADC channels: "——" – sequential registration of phases; "-----" – sequential registration of signals.

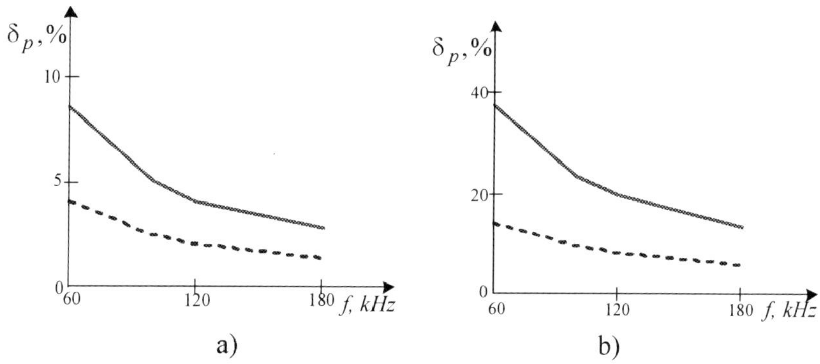

Figure 6.2. The dependence of the relative value of the 2nd harmonic of three-phase active power on the discretization frequency of the ADC at different load angles: a) $\varphi = 30^{\circ}$; b) $\varphi = 80^{\circ}$: "——" – sequential registration of phases; "-----" – sequential registration of signals.

In the analysis of the error of the calculation of the instantaneous active power of three-phase circuits the variable component in the spectrum of P_2 signal was correlated with the constant power P0 for more visual analysis. As the graphs demonstrate, when ADC discretization frequency reduces and power coefficient decreases, component P_2 grows and reaches almost 40% of the constant component of active power at a discretization frequency of 60 kHz on 6 channels and the angle between current and voltage vectors $\varphi = 80^{\circ}$ (Figure 6.3).

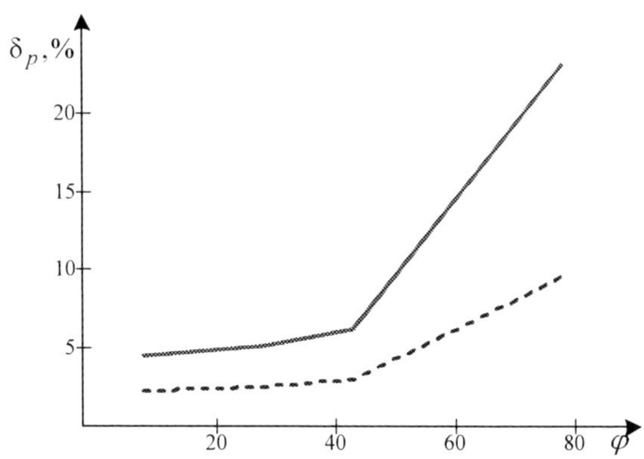

Figure 6.3. The dependence of the relative value of the 2nd harmonic of three-phase active power on the load angle at ADC discretization frequency $f = 100\ kHz$ on 6 channels: "——" – sequential.

It should be noted that the frequency characteristics of current and voltage sensors were not taken into account in the calculations, as they have a fairly wide bandwidth, which for modern sensors is about 30–100 kHz.

The research has shown that the time delay in the serial poll of ADC channels significantly affects the distortion of the results of the calculation of the instantaneous active power of three-phase circuits. In some cases, with "sequential phase registration," the value of the "imaginary" second power harmonic can reach almost 40%, so it is better to conduct a poll of analog channels according to the scheme of "sequential signal registration" as this procedure introduces less error in the calculated power signal. The modeling results indicate that the use of ADCs with a maximum discretization rate of up to 100 kHz without signal correction is unacceptable in the analysis of variable components of three-phase power.

6.2. The Assessment of the Energy Indicators of the Operation of Variable-Frequency Electric Drives on the Basis of Current and Voltage Instantaneous Values

It is known that the effectiveness of VFED operation can be assessed using such energy indicators of IM operation mode as efficiency, power coefficient, components of total power losses, components of instantaneous power of IM. These characteristics can be calculated on the basis of experimentally determined signals of currents and voltages of the phases of the induction motor stator. The application of this method is primarily caused by the fact that, in most cases, in industrial conditions, the phase or line currents and stator voltages of the electric machine are the only parameters that can be easily measured, recorded and assessed.

It is often not possible to conduct the necessary experiments to identify IM parameters during the assessment of the energy parameters of the motor due to the complexity of removing the motor from the process or the absence of necessary equipment, especially in the determination of the stator phase induction. In view of this, another variant of the algorithm for calculating VFED energy performance under the condition of the technological process with minimal information about the parameters of the motor is proposed. During the development of the method, assumptions were made about the possibility of measuring the operation of the motor without load and the possibility of measuring the resistance of the windings of the motor phases after measuring the parameters of the main operating mode in cold and hot state. Besides, to implement the method it is necessary to measure the angular frequency of the rotor, which can be done in a non-contact way, for example, using an optical tachometer.

It is known from [50, 196] that the power on IM shaft can be deter-mined by the equation of power balance:

$$p_2(t) = p_{em}(t)(1 - s(t)) - \Delta p_{f/w} - \Delta p_{ad}(t), \tag{6.3}$$

where

$$p_{em}(t) = p_1(t) - \Delta p_{Cu1}(t) - \Delta p_c(t). \tag{6.4}$$

Then, taking into account (6.4):

$$p_2(t) = p_1(t) - \Delta p_{f/w}(t) - \Delta p_{ad}(t) - \Delta p_{Cu1}(t) - \\ - \Delta p_c(t) - p_1(t)s(t) + s(t)(\Delta p_{Cu1}(t) + \Delta p_c(t)) \quad (6.5)$$

The calculation demonstrates (Table 6.1), that component $s(t)(\Delta p_{Cu1}(t) + \Delta p_c(t))$ can be neglected, as it is insignificant in comparison with other components, especially for medium- and high-power motors, ($\leq 1\%$).

Table 6.1. The error in the calculation of power P_{em}

Parameter \ Motor	$P_н = 30$ kW, $s_н$=0.023	$P_н$= 110 kW $s_н$ = 0.014
$s(t)(\Delta p_{Cu1}(t) + \Delta p_c(t)) / p_1(t)$, %	0.16	0.1

Thus, the power on the shaft:

$$p_2(t) = p_1(t) - \Delta p_{f/w}(t) - \Delta p_{ad}(t) - \Delta p_{Cu1}(t) - \\ - \Delta p_c(t) - p_1(t) \cdot s(t) \quad (6.6)$$

where steel losses $\Delta p_c(t)$, the losses in mechanical components and additional losses $\Delta p_{f/w}(t) - \Delta p_{ad}(t)$ are unknown.

The components of expression (6.6) can be expressed in terms of idle power, because all the power consumed in idle mode is spent on losses in the motor components:

$$p_{id} = \Delta p_{id}(t) = \Delta p_{f/w}(t) + \Delta p_{ad}(t) + \Delta p_{Cu1\_id}(t) + \Delta p_c(t) \quad (6.7)$$

where $\Delta p_{Cu1\_id}(t) = i^2{}_{A\_id}(t) R_A + i^2{}_{B\_id}(t) R_B + i^2{}_{C\_id}(t) R_C$ – electrical losses in the stator copper in idle mode.

We assume that all components of expression (6.7), except for losses in the stator copper, are constant and do not depend on the load. Then, based on the above, and knowing the instantaneous power consumption in idle mode, which is calculated from the instantaneous signals of voltages and currents of the stator of IM operating in idle mode:

$$p_{1id}(t) = u_{A\_id}(t)i_{A\_id}(t) + u_{B\_id}(t)i_{B\_id}(t) + \\ + u_{C\_id}(t)i_{C\_id}(t) \tag{6.8}$$

it is possible to write an expression for IM constant losses that do not depend on the load:

$$\Delta p_k(t) = p_{id}(t) - \Delta p_{Cu1\_id}(t) = \Delta p_{f/w}(t) + \Delta p_{ad}(t) + \\ + \Delta p_c(t) \tag{6.9}$$

Taking into account (6.9) the power on the motor shaft (Figure 6.4) is determined as follows:

$$p_2(t) = p_1(t)(1 - s(t)) - \Delta p_{Cu1\_l}(t) - \Delta p_k(t) \tag{6.10}$$

where $\Delta p_{Cu1\_l}(t) = i^2{}_{A\_l}(t)R_A + i^2{}_{B\_l}(t)R_B + i^2{}_{C\_l}(t)R_C$ – electrical losses in the stator copper in the current operation mode.

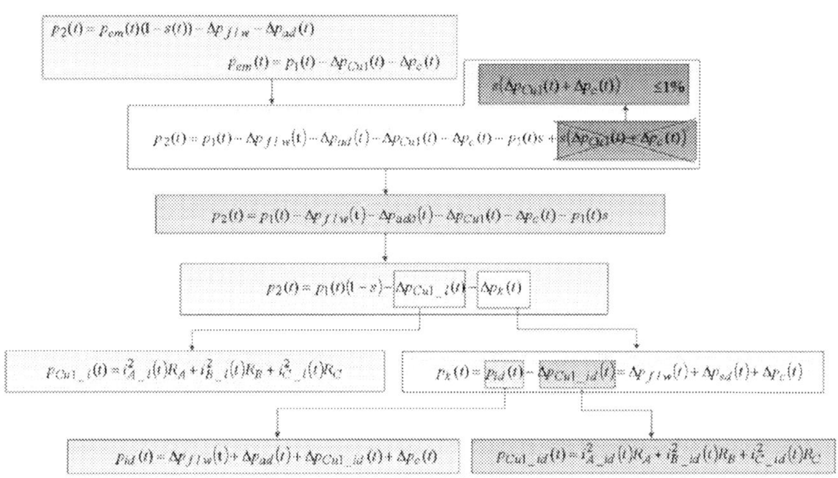

Figure 6.4. The procedure for calculating the components of IM power with minimal information on the parameters of IM EC.

The torque on the motor shaft:

$$T_e(t) = \frac{p_2(t)}{\omega(t)} \tag{6.11}$$

The calculation of these energy parameters showed sufficient accuracy of calculations, for example, the error in calculating the power on IM shaft does not exceed 5%, and for motors with $P_H \geq 100$ kW it does not exceed 1% (Table 6.2).

Table 6.2. The error in calculating power P_B

Parameter \ Motor	P_r = 30 kW, s_r=0.023	P_r = 110 kW s_r = 0.014
The error in calculating power on IM shaft P_2, %	2.1	1.03

The efficiency of one cycle of the electric drive operation, as in the first case, is determined according to expression 6.12.

$$\eta = \frac{\frac{1}{T}\int_t^{t+T} p_2(t)dt}{\frac{1}{T}\int_t^{t+T} p_1(t)dt} \tag{6.12}$$

The algorithm for the indirect determination of energy performance of VFED with minimal information on the parameters of IM EC is presented in Figure 6.5. Its main advantages include the absence of determining all EC parameters, except for the resistance of the stator phases, and conducting a classical experiment of idle speed with a change in supply voltage The implementation of this method is possible provided that: measurements at IM operation in the modes without loading and with rated, or close to it loading are carried out; the measurement of the resistance of the windings of the motor phases is performed after the measurement of the parameters of the main operating mode; the stator phase resistances are measured in cold and heated state.

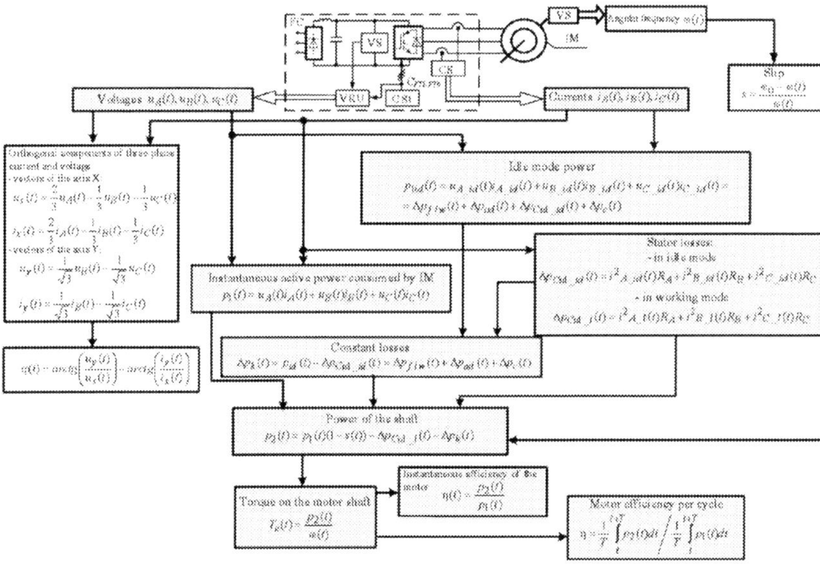

Figure 6.5. The algorithm of the indirect determination of VFED energy indicators with minimal information on the parameters of EC IM.

For the tasks of experimental assessment of the accuracy of the method of determining the power on the shaft, a second version of the research laboratory stand was also developed. It allows performing IM tests under load (Figure 6.6). The stand includes two identical IMs, type AIR80V4U2.The IMs are connected via the dynamic rotary torque sensor ESMNJ0 (torque range – 50Nm, output frequency – 5-15 kHz, accuracy – ±0.5%). An incremental encoder type E40S8-100-3-T-24 is used to control the rotation frequency (100 impulses per revolution).

The Toshiba VFS11 (UZ1) and INVT GD100 (UZ2) FCs were used to power the loaded and loading motors, respectively. UZ1 FC operates in the mode of scalar frequency control and UZ2 FC operates in the mode of vector control with the electromagnetic torque stabilization. Motor $IM1$ operates in the driving mode, and $IM2$ – in a generator mode with a constant torque on the shaft, creating the necessary load for motor $IM1$. Combining the FCs on the DC bus allows the energy regenerated by *IM*2 motor to be transferred via UZ1 FC inverter to motor *IM*1. Thus, the main energy flux circulates along the channel $UZ1-IM1-BT1-IM2-UZ2-UZ1$. In this case, only power losses in frequency converters and motors are consumed from the supply network. To filter the voltage at the output of UZ1 FC, an LC filter is used (L1, C1–C3). The use of LC filter provides total harmonic distortion at a frequency of 50Hz with voltage $THD_U \leq 4\%$ at PWM frequency equal to 8 kHz.

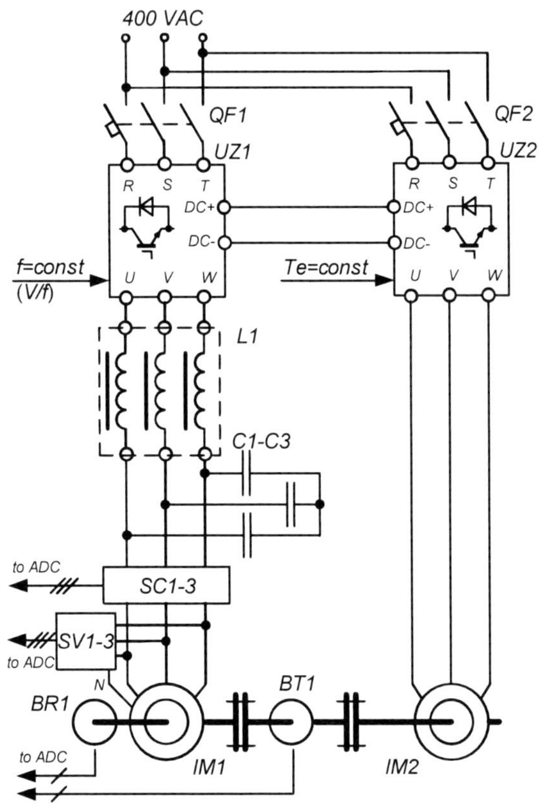

Figure 6.6. Research laboratory stand for testing IM under load.

The currents and voltages in *IM1* circuit were controlled by sensors of current *SC1–SC3* and sensors of voltage *SV1–SV3*, respectively. The signals from the sensors of current (*SC1–SC3*), voltage (*SV1-SV3*), torque (*BT1*) and rotation frequency (*BR1*) were transmitted to an analog-to-digital converter (ADC) for digitization and transfer the data to a computer.

USB module *LCard E14-440* (14 bits, 400 kHz, 32 channels) was used as *ADC*. *LA25-NP* sensors were used as the sensors of current (input – 25 A, output – 25 mA). The sensors of voltage were represented by precision voltage dividers with galvanic isolation based on amplifiers ISO124P.

The signals from the outputs of the rotation frequency (BR1) and torque (BT1) sensors were pulsed frequency signals with maximum frequencies of 2.5 kHz and 15 kHz, respectively. *ADC* module was adjusted for frequency 40 kHz per channel. The recalculation of the pulsed signal into a continuous one was carried out programmatically using the mathematical package Mathcad, in accordance with the algorithm (Figure 6.7), where M – the number of pulse signal averaging periods. The torque signal was recalculated similarly.

```
ni := | n[0,1] ← 0
      | edge ← 0
      | j ← 1
      | for i ∈ 0.. N − 2
      |   if ns[i] = −1 ∧ ns[i+1] = 1
      |     | if edge = 0
      |     |   | t1 ← i·dt
      |     |   | edge ← 1
      |     |   | n[0,0] ← t1
      |     | otherwise
      |     |   | edge ← edge + 1
      |     |   | if edge = Ma + 1
      |     |   |   | t2 ← i·dt
      |     |   |   | n[j,1] ← 1 / ((t2 − t1)/Ma · k)
      |     |   |   | n[j,0] ← t2
      |     |   |   | j ← j + 1
      |     |   |   | edge ← 1
      |     |   |   | t1 ← t2
      | n
```

Figure 6.7. The algorithm for recalculating the impulse signal of the rotation frequency into a continuous one.

The experimental assessment of the accuracy of the method for determining the power on the shaft was performed on a research laboratory stand, which allows the implementation of tests of IM under load (Figure 6.6).

According to (6.10), power $p_2(t)$ on IM shaft was calculated and compared with power $p_{2\,\exp}(t)$ on the shaft that was obtained based on the direct measurement of the torque and rotation frequency:

$$p_{2\,\exp}(t) = T_e(t)\omega(t), \qquad (6.13)$$

where $T_e(t)$ – the torque on IM shaft, obtained experimentally using a torque sensor, $\omega(t)$ – the angular rotation frequency.

Figures 6.8-6.10 contain the results of the experiments at *IM1* frequency start with the start time 0.2 s.

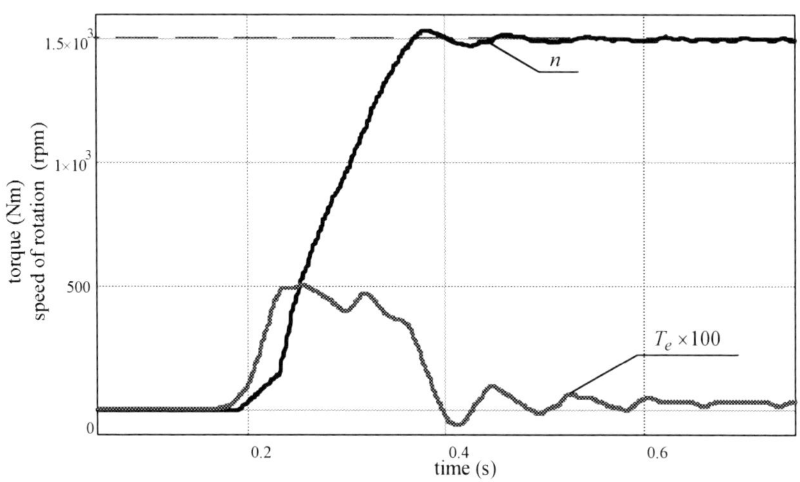

Figure 6.8. The torque on the shaft and IM rotation frequency at a frequency start.

Instantaneous active $p_1(t)$, reactive $q_1(t)$ consumed powers and power $p_2(t)$ on IM shaft were calculated (Figure 6.10) based on the measured current and voltage signals (Figure 6.9).

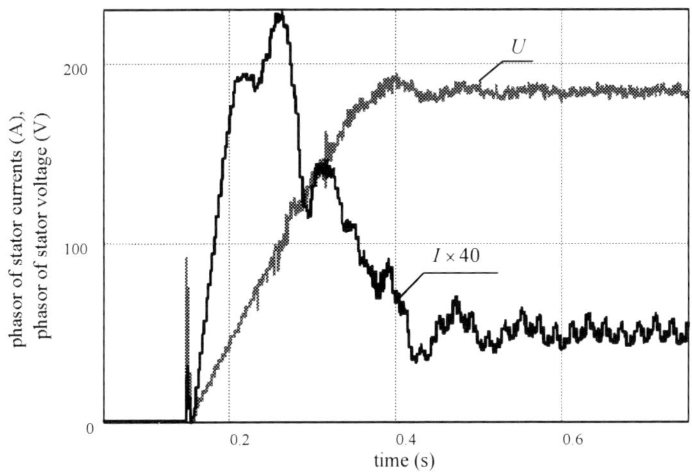

Figure 6.9. The vectors of the stator current and voltage (root-mean-square value) at a frequency start.

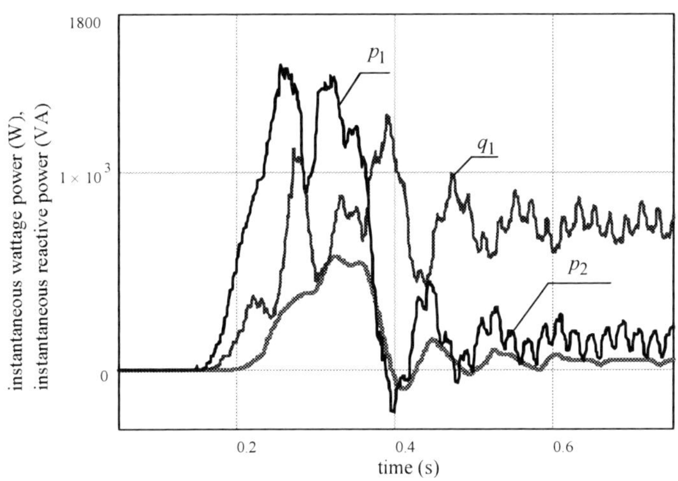

Figure 6.10. The instantaneous active and reactive powers and the power on the shaft at a frequency start.

In another experiment (Figures 6.11-6.14) the loads on IM shaft from the minimal value to 1.1 of the power on the shaft were measured during time $t \approx 0.75$ s. In this case, the electromagnetic time constant of the researched IM was $T_E = 3.5$ ms, and the electromechanical – $T_M = 7$ ms. Due to the small values of IM time constants compared to the load change time, the dynamic components of IM moment were considered insignificant and the accuracy of determining the power on the shaft in this section of the experimental curves was assessed as for the static operation mode.

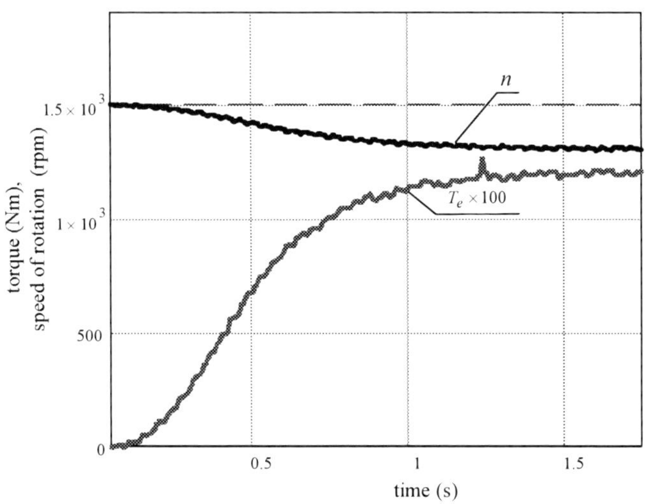

Figure 6.11. The torque on the shaft and IM rotation frequency at increased load.

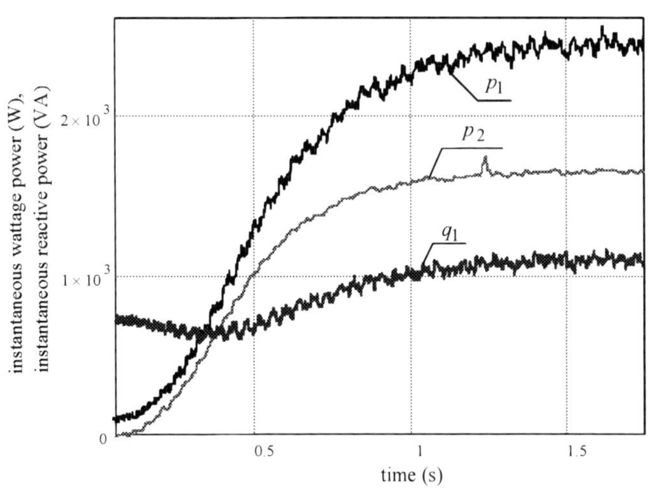

Figure 6.12. The active and reactive powers and the power on the shaft at increased load.

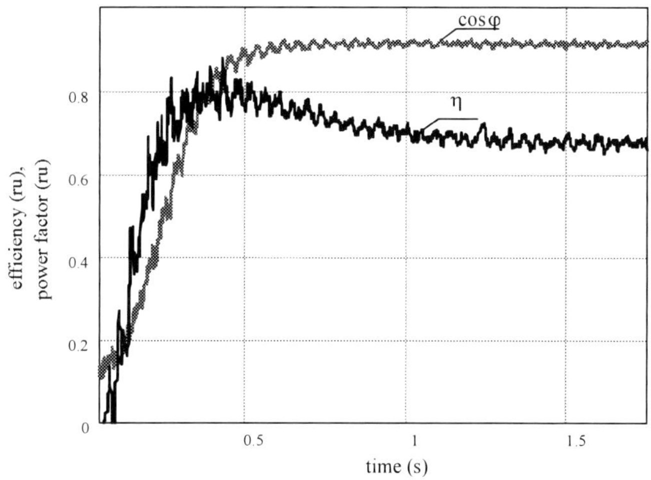

Figure 6.13. The efficiency and IM power coefficient at increased load.

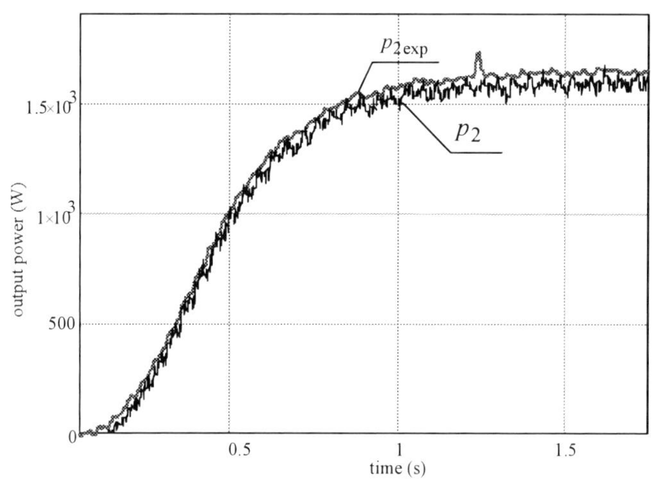

Figure 6.14. The power on IM shaft (calculated $p_{2c}(t)$ and experimental $p_{2\exp}(t)$).

The determination of the power on IM shaft was assessed by comparing the obtained values based on calculation $p_{2c}(t)$ and experimental measurements $p_{2\exp}(t)$:

$$\Delta = \left| \frac{p_{2\exp}(t) - p_2(t)}{p_{2\exp}(t)} \right| \cdot 100\% \tag{6.14}$$

.

The temperature changes in the IM stator resistance or their inaccurate determination are the main source of additional errors in all methods for calculating the power on the shaft and energy indicators. To assess the influence of the accuracy of determining the stator resistance on the accuracy of calculating the power on the IM shaft, a variation of the stator resistances R_s was performed within $\pm 10\%$ (Table 6.3).

Table 6.3. The error in calculating the power on IM shaft

Power on IM1 shaft, r. u.		**0.3**	**0.5**	**0.75**	**1.0**	**1.1**
Error Δ, %	$0.9R_s$	9.36	5.83	2.80	1.52	1.04
	R_s	9.51	6.32	3.81	3.20	3.34
	$1.1R_s$	9.66	6.81	4.81	5.00	5.63

The analysis of the results showed a high accuracy in determining the energy indicators obtained using the indirect method for calculating the power on IM shaft. So, in the load range from 0.5 to 1.1 of the rated power, even when determining the IM stator resistance with an error of $\pm 10\%$, the error of the developed method did not exceed 7%. For the load range from 0.6 to 1.1 the rated power with accurate determination of IM stator resistance, the error of the developed method did not exceed 5%.

6.3. Working out a Method for the Improvement of the Economic Efficiency of the Operation of Variable-Frequency Electric Drive Induction Motors

Taking into account the fact that the main losses of industrial enterprises are associated with the operation of motors with degraded energy performance, the effect of the introduction of the method for determining the optimal period of IMs of variable-frequency EDs by the criterion of minimum costs is calculated according to the following items:

- saving money by reducing the power consumed by the electric drive with the timely replacement of the old motor with a new one;
- saving money by reducing the number of repairs for a specified period;
- saving money by determining the exact service life of the tested motor and timely preventive repairs;
- saving money by reducing the number of repairs and downtime associated with overhauls.

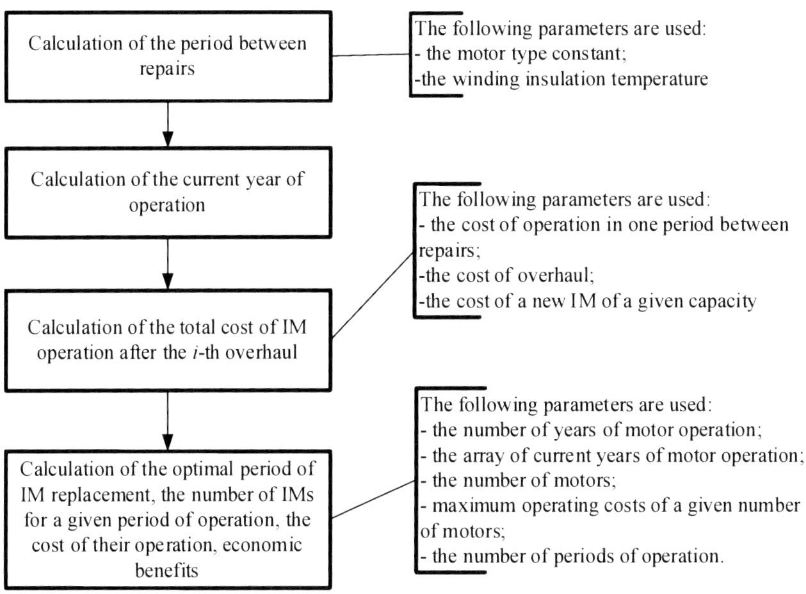

Figure 6.15. The block diagram of the algorithm for determining the optimal period of operation of IMs of variable-frequency ED according to the criterion of minimum cost.

When using the algorithm, the following assumptions are accepted:

- the service life of the electric machine is determined by the well-known rule of "eight" degrees [183];
- the temperature of the windings insulation varies in proportion to the heating losses of the motor;
- the cost of IM operation is calculated for a long-term operation of IM;

- the cost of IM operation is calculated according to the average daily electricity tariff;
- the first overhaul takes place after 5 years of IM operation;
- motor operation is terminated when the period between the repairs is reduced to 2.5 months.

An algorithm for the determination of the optimal period for replacing the motor with a new one was developed using the data on energy losses in IM. The block diagram of this algorithm is presented in Figure 6.15.

In the presented algorithm the first block is the block of input of all set parameters: the rated data of the motor, the cost of one kW of consumed power, the cost of one motor, the cost of one overhaul, the number of working days per year, the number of work shifts, the number of hours per shift, the duration of the work cycle, the rate of increase of losses after each subsequent overhaul.

The next block contains the calculation of the period between repairs according to formula:

$$\Delta\tau(i) = T_0 \cdot e^{-b\cdot\Theta(i)}, \quad i = 1,2..N, \tag{6.15}$$

where $\Theta(i) = \Theta_n \cdot \dfrac{\Delta Pgr(i)}{\Delta Pgr_n}$, $i = 1,2..N$ – the windings insulation temperature after the i-th overhaul [183] (when determining the excess temperature of the windings, you can use the ratio between the rated losses and the rated windings insulation temperature and assume that this factor is constant and does not depend on the mode of operation of the motor); Θ_n – the windings insulation rated temperature determined by IM insulation class; ΔPgr_n – rated heating losses; ΔPgr_i – heating losses after the i-th overhaul; i – the serial number of the overhaul; N – the planned number of overhauls.

The graph demonstrates that the service life of the motor changes according to the exponential law, and when the losses increase by 15%, the service life is four times shorter (Figure 6.16). Therefore, in order to effectively use the company funds, it is necessary to timely replace old motors that have been repeatedly repaired. Thus, after the sixth overhaul, the service life is reduced to 0.8 years.

The initial data for the calculation are given in Table 6.4 (the prices for motors are provided by the company "Electromotor," Kyiv).

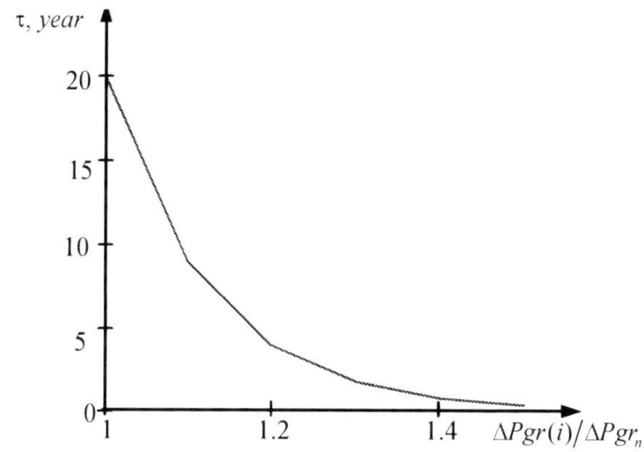

Figure 6.16. The dependence of the residual period of IM operation on the relative heating losses.

Table 6.4. The initial data for the calculation

No.	Indicators and the units of their measurement	Symbol	Indicator value
1	Number of calendar days per year, days	Q_D	365
2	Number of working days per year, days	Q_w	250
3	Duration of a shift, hour	T_S	8
4	Number of shifts, pcs	Q_S	2
5	Motor cost, UAH:	C_{IM}	
	1.1kW		460
	11kW		2570
	110kW		34400
6	Cost of an overhaul, UAH	C_o	
	1.1kW		200
	11kW		1200
	110kW		11000
7	Coefficient taking into account transport costs	α_t	1.05
8	Coefficient taking into account installation costs	α_c	1.1
9	Cost of 1 kWh of electricity, UAH	C_e	0.71

The current year of IM operation is determined taking into account that the first overhaul is carried out after 5 years of its use, and its operation becomes impractical when the period between repairs reduces to 3 months ($\tau(i_{end})$):

$$\tau(i) = 5 + \sum_{i=0}^{\Delta\tau(i) \geq 0.25} \Delta\tau(i), \quad i = 1.2..N \tag{6.16}$$

The cost of operation of one IM after the i-th overhaul: the cost of a new motor, the number of overhauls, the cost of one such overhaul and the payment for electricity consumed by the motor to cover energy losses, taking into account the increase in losses after each subsequent overhaul:

$$\Sigma(i) = C_{AD} \cdot \alpha_t \cdot \alpha_c + C_k \cdot i + \sum_{1}^{i} i \cdot \Delta\tau(i) \cdot H, \quad i = 1,2..N, \quad H = C_e \cdot \Delta p \cdot \Delta P_\Sigma \cdot Q_p \cdot t_c \cdot q_c, \tag{6.17}$$

where Δp – the increase in total heating losses of IM after each overhaul, r.u.; ΔP_Σ – the total heating losses of the motor in rated mode, kW.

Based on this, the cost of operation of n motors is $\Sigma_n = n \cdot \Sigma(i_{end})$ for the period of time $n \cdot \tau(i)$ ($n = 1,2..5$) according to the standards.

The last block of the algorithm is the block of deciding on the optimal period for replacing IM with a new one, the number of motors per cycle, operating costs and economic benefits. This procedure is performed by constantly comparing the cost of operation of a given amount of IM n and the recommended amount m. Thus, the moment of time is determined when it is more expedient to put into operation a new motor at a minimum cost for the set period $n \cdot \tau(i_{end})$.

$$\Sigma(i) = \Sigma(i_{end}) - \Sigma(i_{opt}) \to \max, \tag{6.18}$$

on conditions:

$$n \cdot \tau(i_{end}) = m \cdot \tau(i_{opt}), \quad m > n, \text{ at } m, n = 1,2..5, \quad i_{opt} < i_{end} \text{ at } i = 1,2..N. \tag{6.19}$$

Induction motors with rated data given in Table 6.5 were the object of the calculation.

To obtain data, the calculation is performed as follows. Initially, the rated service life of IM is calculated separately for each motor, which depends on the insulation class of its windings. Then the developed algorithm is used to determine the optimum term of replacement of the motor taking into account the increase in total losses in IM after each overhaul and, as a result, the reduction of the period between repairs. The calculations were performed for the following cases: $\Delta p = 0.025$, $\Delta p = 0.05$, $\Delta p = 0.075$ and relative load torque $T_e^* = 0.8, 0.9, 1$. This assumption can be made on the basis of the analysis of experimental and calculated data obtained by assessing the operating parameters of the repaired induction motors [31, 251].

Table 6.5. The published data of the researched IMs

Power, kW	**1.1**	**11**	**110**
Stator current, A	2.5	22	199
Rotation frequency, rev/min	2811	1458	982
Efficiency, %	0.775	0.875	0.93
Power coefficient	0.87	0.87	0.9

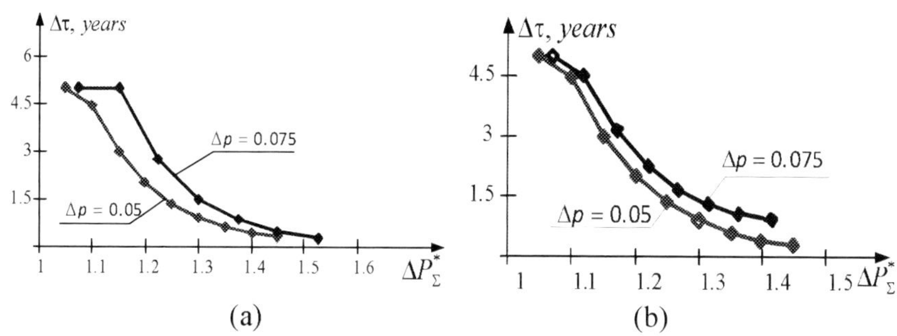

Figure 6.17. The dependence of IM period between repairs on relative total losses: a) for IM with $P_r = 110$ kW, b) for IM with $P_r = 1.1$ kW.

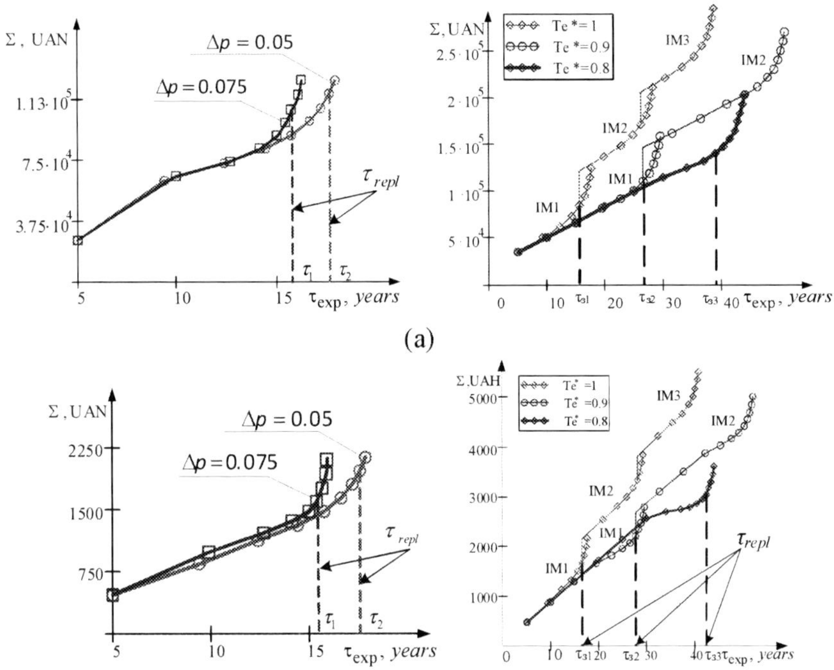

Figure 6.18. The dependence of the operation costs on the current year of the motor operation: a) for IM with $P_r = 110$ kW, b) for IM with $P_r = 1.1$ kW.

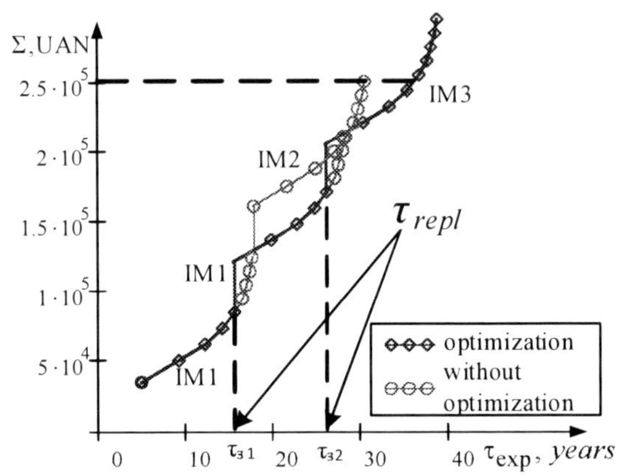

Figure 6.19. The motor operation period calculated with the use of the algorithm (optimization) and according to the published data (without optimization).

The above graphs demonstrate that the economic efficiency of IM operation sharply decreases after several overhauls. Figure 6.18 shows how the frequency of overhauls increases with the gradual growth of total losses in the motor, which, in turn, causes a sharp increase in operating costs at some point in time. Thus, Figure 6.19 illustrates that with timely motor replacement and constant operating costs, IM service life increases by almost 10 years. It should be noted that the shapes of the constructed curves are almost the same for motors of different power, which makes the algorithm universal for different types of IM.

The presented algorithm allows determining the optimal service life of induction motors of variable-frequency ED by the criterion of minimum costs. Its application reduces the expenditures of industrial enterprises associated with the operation of electromechanical equipment with degraded energy performance [252]. To increase the reliability of the obtained results, the proposed algorithm is recommended for the use together with the method for calculating IM loss based on the instantaneous values of the signals of the stator currents and voltages. Improving the accuracy of the algorithm is possible with the use of reliable methods for determining losses in IM units, as well as taking into account the actual modes of operation.

The economic effect of the algorithm increases significantly in the presence of a large fleet of electric motors of different capacities at industrial enterprises.

Conclusion

1. The quantitative assessment of the error of the calculation of the active instantaneous power of three-phase circuits at the serial poll of ADC voltage and current channels has been carried out. It has been shown that in some cases, with the "sequential registration of phases," the value of the "imaginary" second power harmonic can reach almost 40%. Recommendations on the procedure for polling the current and voltage channels of the phases of IM stator aiming at making a smaller error in the calculated instantaneous power signal have been formulated. It has been shown that the use of ADCs with a maximum discretization frequency of up to 100 kHz without signal correction is unacceptable in the problems of the analysis and compensation for the variable components of three-phase power.

2. The structure and parameters of the test complex have been offered, by means of which the method of the assessment of the energy parameters of VFED with an asymmetric IM was experimentally verified on the basis of instantaneous signals of currents and voltages of the stator phases.
3. The method of determining the energy characteristics of VFED operation has been experimentally tested on the basis of instantaneous values of IM currents and voltages. The research of AIR80V4U2 IM has shown that the presence of 14.03% asymmetry of the winding of one of the stator phases results in more than 7% asymmetry of the stator current in the load mode and in a variable component of the electromagnetic torque in 16.74%.
4. It has been shown that the application of the proposed algorithm for determining the optimal service life of induction motors of variable-frequency electric drives, based on real data on the technical condition of IM, by the criterion of minimum costs reduces the industrial enterprises costs associated with the use of IM with deteriorated energy characteristics due to timely overhauls and current repairs or replacements. The economic effect of the algorithm increases significantly in the presence of a large fleet of electric motors of different capacities at industrial enterprises.
5. The experimental assessment of the accuracy of the method for determining the power on the shaft of an induction motor has shown the acceptable accuracy even in conditions of determining the resistance of the stator with a significant error. The error of indirect determination of power on the shaft did not exceed 5% for the researched induction motor at loads from 60% to 110% of the rated power. The conducted experimental research substantiates the use of the proposed method in tests of new and repaired motors, as well as in monitoring, diagnostics, fault-tolerant control of electric drives with frequency control.

Conclusion

1. Based on the analysis of the design features and the statistics of induction motor damages, it has been shown that the most common causes of variable power and electromagnetic torque include magnetic and electrical asymmetry of stator windings, which is the result of electrical damage such as turn-to-turn short circuits, break of the parallel section of the stator phase winding, break of one elementary conductor in the winding.
2. It has been experimentally proved that the mathematical model of an induction motor in a three-phase coordinate system taking into account the degree of asymmetry of the stator windings allows adequate representation of the electromagnetic and electromechanical processes occurring in induction motors motors. Even with significant asymmetry about 10–15% the calculation error is up to 5% for idle mode and up to 1.5% in the operation with rated load.
3. The structure of VFED control system has been proposed; it uses the cross-vector instantaneous power theory, which makes it possible, without changes in the power part of the frequency converter, to compensate for the influence of the asymmetry of the induction motor stator windings on the mode of operation of the electric drive by eliminating the variable components of the consumed three-phase power or electromagnetic torque.
4. VFED mathematical modeling has shown that the asymmetry of the induction motor stator windings causes the redistribution of currents in the stator phases, resulting in local overheating of the stator windings, with asymmetry of about 2–10%, and increasing the copper losses in individual windings by more than 10–30%. Another negative result of asymmetric current loading of induction motor phases consists in the redistribution of currents in individual power semiconductor switches of the inverter and, as a consequence, increased losses and local overheating of these switches.

5. It has been shown that the level of compensation for the variable power component and the electromagnetic torque of the induction motor depends on the value of the gain coefficient in the direct channel of the control system, the load level of the motor and the degree of stator phase asymmetry. The gain coefficient dependence on the load level and the specified level of compensation for the variable component, obtained by creating a regression model, enables building an adaptive control system to compensate for the impact of asymmetry of the induction motor stator windings.
6. The results of mathematical modeling of the VFED system with the compensation for the influence of the asymmetry of the stator windings on the mode of operation of the electric drive by eliminating the variable component of the electromagnetic moment have shown that the proposed control system allows reducing the specified component to 1–2% with the asymmetry of the stator windings in the range of 5–15%, which significantly improves the vibration state of IM, but does not significantly reduce the thermal overload of the individual phases of the motor and the corresponding SVI power switches.
7. It has been shown that adjusting the control system of the variable-frequency electric drive when compensating for the variable component of the consumed three-phase power with the asymmetry of the windings in the range of 5–15%, reduces the thermal overload of individual stator windings and corresponding power semiconductor switches of the autonomous voltage invertor by more than 50% due to the redistribution of currents. The latter, in turn, increases the service life of the motor insulation by almost 6 times, compared with the operation without compensation for the asymmetry of the windings. However, the use of this setting of the compensation system reduces the efficiency of eliminating the variable component of the electromagnetic moment.
8. Methods and systems for fault-tolerant frequency vector control of the three-phase induction motors in case of damage in the stator power circuit have been developed. These methods and systems are based on the introduction of additional control variables in various control circuits, which creates an asymmetric stator voltage compensating for the asymmetry caused by a damage to the induction motor. The choice of the fault-tolerant frequency vector control system from the proposed ones is determined by the tasks that are set

in the event of damage in the power circuit of the induction motor stator: the compensation for the variable component of the electromagnetic torque, the compensation for the variable component of the three-phase active power, the reduction of thermal overload of separate phases of the induction motor and power switches of the frequency converter.

9. The fault-tolerant frequency vector control system with the function of compensation for a variable component of the electromagnetic torque of the induction motor has been developed. In this control system, the variable component of the electromagnetic torque is located at the input of the regulator of the torque-generating component of the stator current in the speed control channel. The proposed method of the induction motor fault-tolerant control allows the reduction of the variable component of the motor electromagnetic torque to an acceptable level with a slight reduction in variable components of power consumption and losses in the power part of the variable-frequency electric drive.
10. The fault-tolerant frequency vector control system with a function of compensation for the variable component of active consumed power of induction motor, based on the modified p-q power theory, has been developed. The reseach results demonstrate that the use of the proposed fault-tolerant control system with the function of compensation for the variable component of the three-phase active power can significantly reduce the thermal overload of the individual phases of the induction motor and power switches of the frequency converter. However, there is a slight increase in the variable component of the electromagnetic torque of the induction motor.
11. The fault-tolerant frequency vector control system with separate regulation of flux linkage circuits and active current component has been created for each phase of induction motor separately. This makes it possible to adjust the regulators of phase flux linkage according to the asymmetry of electromagnetic parameters of the corresponding phase of the motor and to correct the operating modes of the electric drive by the change of reference signals of the phase linkage. At the same time reduction of flux linkage of the damaged phase allows decreasing a variable component of the motor electromagnetic torque to an admissible level and reducing the thermal overload of separate phases of the motor and semiconductor keys of frequency converters.

12. A method for identifying the electromagnetic parameters of an induction motor as part of a variable-frequency electric drive has been proposed. The interphase parameters are determined by setting test effects on the stator windings in the form of sinusoidal low-frequency voltages in the pre-start-up period. Their further recalculation allows determining the phase parameters of the equivalent circuit and the levels of asymmetry of the stator windings of the motor to set up a separate vector control system. The advantages of the proposed method include the simplification of the mathematical apparatus of identification and sufficient accuracy in determining the electromagnetic parameters through the use of iterative algorithms based on the comparison of experimental and calculated current signals.
13. The results of experimental research have shown the efficiency of the developed methods for determining the initial approximations and direct electromagnetic parameters of the induction motor equivalent circuit, the errors of which do not exceed 2% for resistance, 1.5% – for leakage inductances, 4% – for the magnetization circuit inductance. The application of the developed methods in industrial specimens of variable-frequency electric drives with the developed measuring system makes it possible to solve the problem of identification of electromagnetic parameters of the equivalent circuit of an induction motor in the pre-start-up period for each phase separately.
14. An algorithm for calculating the maximum recommended load level of an induction motor at different degrees of asymmetry has been developed from the point of view of ensuring the allowable insulation temperature of the windings when compensation is impossible. The application of this approach in the conditions of technological process that enables keeping IM loading at a certain level makes it possible to improve essentially the operation conditions of the motor and to increase its service life.
15. The developed method of indirect determination of power on the shaft and energy performance of VFED with IM allows the assessment of the energy efficiency of VFED in conditions of motor asymmetry and changes in its modes by components of total losses, power coefficient, efficiency, instantaneous stator, rotor and motor power conversion factor. The advantage of the developed method consists in the fact that it reduces the time spent on data analysis compared with the

methods that involve the decomposition of current and voltage signals into harmonic components. The proposed method allows determining the energy performance of ED operation without removing the IM from the process, using minimal information about the object of research, in the conditions of practical impossibility of use of the power equipment for creation of special operating modes and formation of special test influences.

16. The proposed algorithm for determining the optimal period of operation of induction motors of variable-frequency electric drives using the results of indirect determination of their energy performance reduces the industrial enterprise costs associated with the operation of induction motors with degraded energy performance by almost 20% due to timely overhauls, repairs or changes.

References

[1] Kolpakov A. I. Perspective for the electric drive development. *Power electronics,* 2004, № 1, pp. 46–48.

[2] Pankratov V. V. The development trend of general industrial AC drives based on modern power electronics devices. *Power intelligent electronics,* 2005, № 2, pp. 46–52.

[3] Ilyinsky N. F. Adjustable electric drive. Energy and resource saving. *Drive engineering,* 1997, № 3, pp. 21–23.

[4] Sandler A. S., Sarbatov R. S. *Automatic frequency control for induction motors.* – Moscow: "Energy," 1974, 144 p.

[5] Bichai V.H., Piza D. M., Potapenko E. E., Potapenko E.M. State, trends and problems in the field of control methods for induction motors. *Radio electronics, informatics, control,* 2001, № 1, pp. 138–144.

[6] Zabrodin Yu. S. *Industrial electronics: textbook for universities.* – M.: Vysshaia shkola, 1982, 496 p.

[7] Usoltsev A. A. *Frequency control of induction motors: textbook.* – St. Petersburg: Saint Petersburg State University of Information Technologies, Mechanics and Optics, 2006. – 94 p.

[8] Sokolovskiy G. G. *AC electric drives with frequency control: textbook for university students.* – Moscow: Akademiia, 2006. – 272 p.

[9] Blaschke F. The Principle of Field Orientation as applied to the New Transvector Closed Loop Control System in a PWM Inverter Induction Motor Drive. *Siemens Review,* 1972, vol. 39, no. 5, pp. 217–220.

[10] Jezernik K. Robust Direct Torque and Flux Vector Control of Induction motor. *Proceeding of IECON '98.* Germany, Sept., 1998, vol. 2, pp. 667–672.

[11] Wang Wen-Jieh, Wang Chun-Chieh A New Composite Adaptive Speed Controller for Induction Motor Based on Feedback Linearization. *IEEE Transaction on Energy Conversion,* 1998, vol. 13, № 1, pp. 1–6.

[12] Franceschini G., Piazzi A., Tassoni C. Genetic Algorithm Approach to Design Flux Observer for Induction Servo Motors. *Proceeding of IECON '94,* 1994, vol. 3, pp. 2132–2136.

[13] Chen T. C., Liaw C. Y. Design of a Neural Fuzzy Controller for Induction Motor Speed Control. *Proceeding of IECON '94,* 1994, vol. 1, pp. 611–616.

[14] Lopez-Toribio C. J., Patton R. J., Daley S. Fuzzy Fault-Tolerant Control of a Rail Traction Sytem. *Proceeding of the 14th IFAC World Congr, 1999,* 1999, pp. 3–4.

[15] Krasova S. A., Utkin V. A. Cascade synthesis of asymptotic state observers of an asynchronous sensorless electric drive. *Proceedings of the International Conference "System Identification and Control Problems "(SICPRO '2000),* Moskow: Institute of management problems, 2000, pp. 24-28.

[16] Takahashi L., T.Noguchi T. A new quick response and high efficiency strategy of induction motor. *IEEE Transactions on Industry Applications,* 1986, vol. IA-22, no. 5, pp. 820-827.

[17] Krzeminski Z. Application of Observer System to Nonlinearly Controlled Induction Motor Fed by Voltage Source Inverter. *Proceeding of IECON '94.* Italy, Bologna. Sept., 1994, vol. 1, pp. 311–315.

[18] Janda Z., Jankovic' M., Bebic' J., Vukosavich' S., Vuckovich' vol. The Realization of a Novel Speed-Sensorless Induction Motor Drive. *Proceeding of IECON '94,* 1994, V. 3, pp. 1621–1626.

[19] Birou I. Robust speed control of a sensorless driving system with field-oriented induction machine. *Proceedings of the 9th International Conference on Optimization of Electrical and Electronic Equipment OPTIM'2004,* Brasov, Romania, May 20–21, 2004, pp. 91–96.

[20] Braslavski I. YA., Ishmatov Z. Sh., Poliakov V. N. *Energy-saving asynchronous electric drive: textbook for university students* [ed. by Braslavski I. YA.]. – Moscow: Akademiia, 2004. – 256 p.

[21] Zakladnoi O. M., Prakhovnyk A. V., Solovei A. I. *Energy saving by means of an industrial electric drive: tutorial.* – Kyiv: Kondor, 2005. – 408 p.

[22] Ilyinsky N. F., Rozhankovsky Yu. V., Gornov A. O. *Energy saving in electric drive.* – Moscow: Vysshaia shkola, 1989. – 122 p.

[23] Frederick L. H. Drive system efficiency improvements. *Proceedings of the XIX International Conference on Electrical Machines,* September 6–8, 2010, p. 6.

[24] Poliakov V. N., Shreyner R. T. *Extreme control of electric motors.* – Ekaterinburg, 2006. – 420 p.

[25] Sandler A. S., Sarbatov R. S. *Automatic frequency control of induction motors.* – Moskow: Energy, 1974. – 328 p.

[26] Casadei D., Mengoni M., Serra G., Tani A., Zarri L., Cabanas M. F. Energy–efficient control of induction motors for automotive applications. *Proceedings of the XIX International Conference on Electrical Machines,* September 6–8, 2010, p. 6.

[27] Kohler J. L., Sottile J. , Trutt F. C. Condition based maintenance of electrical machines. *Proceedings of the IEEE Industry Applications Conference, 34th IAS Annual Meeting,* Phoenix, 1999, pp. 205–211.

[28] Grebchenko N. V., Polkovnichenko D. V. Experimental researches of steady-state operating modes of induction motors with the presence of defects in them. *Bulletin of scientific works of DonSTU.* Series: Electrical engineering and energy, Donetsk: DonSTU, 2000, vol. 17, pp. 110–114.

[29] Kokko V. I. J. Electrical ageing in lifetime estimation of hydroelectric generator stator windings. *Proceedings of the XIX International Conference on Electrical Machines,* September 6–8, 2010, p. 6.

[30] Zhao X., Corda J. , Mei T. X. Analysis of the effect of asymmetrical phase parameters in a 3-phase permanent magnet synchronous motor. *Proceedings of the XIX International Conference on Electrical Machines,* September 6–8, 2010, p. 6.

[31] Prus V. V., Kiryanov O. F., Geraskin A. S. Estimation of operational parameters of repaired induction motors. *Bulletin of KSPU. Transactions of KSPU, 2001,* Iss. 1/2002 (12), pp. 174–179.

[32] Bruskin D. E. *Electrical machines.* – Moskow: Vysshaia shkola, 1987. – 319 p.

[33] *Design of electrical machines*/edited by Kopylov I. P. – Moskow: Energy, 1980. – 495 p.

[34] Voldek A. I. *Electrical machines.* – Leningrad: Energy, 1974. – 840 p.

[35] Kokorev A. S. *Handbook for a young electrical machine wrapper.* – Moskow: Vysshaia shkola, 1985. – 207 p.

[36] Kopylov I. P. and etc. *Design of electrical machines: Textbook for universities.* 3rd ed., rev. and additional. – Moskow: Vysshaia shkola, 2002. – 757 p.

[37] Schoen R. R., Lin B. K., Habetler T. G. , Schlag J. H., Farag S. An unsupervised, on-line system for induction motor fault detection using stator current monitoring. *Industry Applications Society Annual Meeting, 1994, IEEE Conference Record,* 2–6 October, 1994, pp. 103–109.

[38] Toliyat H. A., Lipo T. A. Transient analysis of cage induction machines under stator, rotor bar and end ring faults. *IEEE Transactions on Energy Conversion,* 1995, vol. 10, № 2, pp. 241–247.

[39] Fedorov M. M., Dennik V. F., Chornous V. P. Protection of AC motors from unbalanced operating modes in case of breakage of parallel branches in the stator winding. *Bulletin of scientific works of DonSTU,* Donetsk: DonSTU, 2001, vol. 18, pp. 130–132.

[40] Fedorov M. M., Tkachenko A. A. Thermal state of induction motors in the event of a break in the parallel branch of the stator phase winding. *Bulletin of Kremenchuk Mykhailo Ostrohradskyi State University,* 2009, issue 4 (57), part 1, pp. 51–54.

[41] Morin R., Novak J. P. , Bartnikas R. , Ros R. Analysis of in-service aged stator bars. *IEEE transactions on Energy Conversion,* 1995, vol. 10, no. 4, pp. 645–652.

[42] *The effect of repair/rewinding on motor efficiency.* Report of the electrical apparatus service assosiation, 2003, p. 26.

[43] Vaskovsiy Yu. V., Geraskin A. A. Mathematical modeling of electromagnetic fields in a squirrel-cage induction motor with a damaged rotor winding. *Technical electrodynamics,* 2010, no. 2, pp. 56–61.

[44] Gemke R. G. *Malfunctions of electrical machines* [edited by Umantsev R.B.]. – Leningrad: Energoatomizdat, 1989. – 336 p.

[45] Rodkin D. I. On the evaluation of energy process indicators using the provisions of instantaneous power. *Bulletin of NTU "KhPI."* Collection of scientific works. Series: Problems of automated electric drive. Theory and practice, 2013, № 36(1009), pp. 366–370.

[46] Sivokobylenko V. F., Tkachenko S. N. Influence of rotor heating temperature on operating and starting characteristics of deep-groove induction motor. *Bulletin of KSPU. Transactions of KSPU,* 2007, issue 3 (44), Part 2, pp. 66–71.

[47] *Rotating electrical machines. Induction motors.* Test methods: GOST 7217–87. [Acting from 25.03.1987]. – Moscow: IPK Standards Publishing House, 1987. – 28 p.

[48] *Rotating electrical machines. Methods for determining losses and efficiency:* GOST 25941–83. – [Acting from 01.01.84]. – Moskow: IPK Standards Publishing House, 1984. – 30 p.

[49] Bruskin D. E. *Electrical machines.* – Moscow: Vysshaia shkola, 1987. – 319 p.

[50] Voldek A. I., Popov V. V. *Electrical machines.* AC Machines: A Textbook for Universities. – St. Petersburg: Piter, 2008. – 350 p.

[51] Postnikov I. M. *Generalized theory and transient processes of electrical machines.* – Moscow: Vysshaia shkola, 1975. – 319 p.

[52] Korenkova T., Kovalchuk V. and Qawaqzeh M. Z., The Assessment of the Electrohydraulic Complex Power Controllability in the Event of an Emergency Shutdown of the Power Supply. *2020 IEEE Problems of Automated Electrodrive. Theory and Practice (PAEP),* 2020, pp. 1-6, doi: 10.1109/PAEP 49887.2020.9240854.

[53] Zinovyev G. S. *Direct methods for calculating the energy indicators of valve converters.* – Novosibirsk, 1990. – 219 p.

[54] Bacher J., Waldhart F. Efficiency determination of standard asynchronous machines from Start–Up data. *Proceedings of the XIX International Conference on Electrical Machines,* September 6–8, 2010, p. 6.

[55] Kovalchuk V., Korenkova T. and Almashakbeh A. S., Electrohydraulic Complex Parameters Determination Based on the Energy Balance Equations. *2020 IEEE Problems of Automated Electrodrive. Theory and Practice (PAEP),* 2020, pp. 1-6, doi: 10.1109/PAEP49887.2020.9240819.

[56] Grebchenko N. V., Polkovnichenko D. V. Experimental researches of steady-state operating modes of induction motors in the presence of defects in them. *Bulletin of scientific works of DonSTU.* Series: Electrical Engineering and Energy, Donetsk: DonSTU, 2000, vol. 17, pp. 110–114.

[57] Zhezhelenko I. V., Rabinovich M. L., Bozhko V. M. *Power quality in industrial plants.* – Kyiv: Technics, 1981. – 160 p.

[58] Chornyi O. P., Kalinov A. P., Mamchur D. G. Application of energy conversion quality indicators to assess the state and reliability of electromechanical systems. *Bulletin of scientific works of DSTU. Thematic issue "Problems of automated electric drive. Theory and practice,"* Dniprodzerzhynsk: DSTU, 2007, pp. 519–524.

[59] Hamed Shahsavari Alavije and Mahdi Akhbari, Investigation of induction motor drive behavior in low-cost fault tolerant control for electric vehicles. *5th International Power Engineering and Optimization Conference (PEOCO),* 6-7 June 2011, pp. 176 – 181.

[60] Campos D.-Delgado, Espinoza-Trejo D., Palacios E. Fault-tolerant control in variable speed drives: a survey. *IET Electric Power Appl.,* March 2008, vol. 2, no. 2, pp. 121–134.

[61] Cusido J., Romeral L., Ortega J. A., Garcia A., Riba J. Signal injection as a fault detection technique. *Sensors,* March 2011, vol. 11, pp. 3356–3380.

[62] Zagirnyak M., Mamchur D., Kalinov A. Comparison of induction motor diagnostic methods based on spectra analysis of current and instantaneous power signals. *Przeglad Elektrotechniczny* (*Electrical Review*), issue 12b/2012, pp. 221–224. ISSN 0033-2097

[63] Cheng Siwei, Zhang Pinjia, Habetler T. G. An Impedance Identification Approach to Sensitive Detection and Location of Stator Turn-to-Turn Faults in a Closed-Loop Multiple-Motor Drive. *IEEE Transactions on Industrial Electronics,* May 2011, vol. 58, no. 5, pp. 1545–1554.

[64] Yun Jangho, Lee Kwanghwan, Lee Kwang-Woon, Lee Sang Bin, Yoo Ji-Yoon Detection and Classification of Stator Turn Faults and High-Resistance Electrical Connections for Induction Machines. *IEEE Transactions on Industry Applications,* March-april 2009, vol. 45, issue: 2, pp. 666–675.

[65] Yun J., Cho J., Lee S. B., Yoo J.-Y. Online detection of highresistance connections in the incoming electrical circuit for induction motors. *IEEE Trans. Ind. Appl.,* Mar./Apr. 2009, vol. 45, no. 2, pp. 694–702.

[66] Barrera Pablo, Guillermo M., Bossio R., Solsona Jorge A. High-Resistance Connection Detection in Induction Motor Drives Using Signal Injection. *IEEE Transactions on Industrial Electronics,* July 2014, vol. 61, issue 7, pp. 3563–3573.

[67] Nussbaumer Peter, Zoeller Clemens, Wolbank Thomas M., Vogelsberger Markus A. Transient distribution of voltages in induction machine stator windings resulting from switching of power electronics. *39th Annual Conference of the IEEE Industrial Electronics Society, IECON 2013,* 10-13 Nov. 2013, pp. 3189–3194.

[68] Das Santanu, Purkait P., Chakravorti S. Separating induction Motor Current Signature for stator winding faults from that due to supply voltage unbalances. *1st International Conference on Power and Energy in NERIST (ICPEN),* 28-29 Dec. 2012, pp. 1–6.

[69] Ethni S. A., Gadoue S. M., Zahawi B. Inter-turn short circuit stator fault identification for induction machines using computational intelligence algorithms. *IEEE International Conference on Industrial Technology,* March 2015, pp. 757–762.

[70] Grubic Stefan, Aller Jose M., Lu Bin, Habetler Thomas G. A Survey on Testing and Monitoring Methods for Stator Insulation Systems of Low-Voltage Induction Machines Focusing on Turn Insulation Problems. *IEEE Transactions on Industrial Electronics, December 2008,* vol. 55, no. 12, pp. 4127–4136.

[71] Bonnett A. H., Soukup G.C. Cause and analysis of stator and rotor failures in three-phase squirrel-cage induction motors. *IEEE Transactions on Industry Applications,* July/Aug. 1992, vol. 28, no. 4, pp. 921–937.

[72] Goldberg O. D., Helemskaya S. P. *Reliability of electrical machines: a textbook for students of higher education:* edited by Goldberg O. D. – Moscow: Publishing Center "Academy," 2010. – 288 p.

[73] Kyznetzov N. L. *Reliability of electrical machines: textbook for universities.* – Moscow: MPEI Publishing House, 2006. – 432 p.

[74] Stone G. C., Boulter E. A., Culbert I., Dhirani H. *Electrical Insulation for Rotating Machines: Design, Evaluation, Aging, Testing, and Repair.* – NJ: Wiley-IEEE Press, 2004. – 631 p.

[75] Goldberg O. D., Sviridenko I. S. *Engineering design and SAD of electrical machines: a textbook for students of higher education:* edited by Goldberg O.D. – Moscow: Publishing Center "Academy," 2008. – 560 p.

[76] Kotelenetz N. F., Akimov N. A., Antonov M. V. *Testing, operation and repair of electrical machines.* – Moscow: Publishing Center "Academy," 2003. – 384 p.

[77] Yeh Chia Chou, Povinelli Richard J., Mirafzal Behrooz, Demerdash Nabeel A. O. Diagnosis of Stator Winding Inter-Turn Shorts in Induction Motors Fed by PWM-Inverter Drive Systems Using a Time-Series Data Mining Technique. *International Conference on Power System Technology,* Singapore, 2004, pp. 854–860.

[78] Pinjia Z., Yi D., Thomas G. H., Bin L. A Survey of Condition Monitoring and Protection Methods for Medium-Voltage Induction Motors. *IEEE Trans. on Ind. Appl.,* 2011, vol. 47, pp. 34–46.

[79] Siddique A., Yadava G. S., Singh B. A review of stator fault monitoring techniques of induction motors. *IEEE Transactions on Energy Conversion,* March 2005, vol. 20, no. 1, pp. 106–114.

[80] Lietal M. Effects of Repetitive Pulse Voltages on Surface temperature Increase at End Corona Protection Region of HV Motors. *Insucon, Birmingham,* May 2006, pp. 105–108.

[81] Zhang W., Xu D., Enjeti P.N., Li H., Hawke J.T., Krishnamoorthy H.S. Survey on Fault-Tolerant Techniques for Power Electronic Converters. *IEEE Trans. on Power Electronics,* August 2014, vol. 29, pp. 6319–6331.

[82] Wu Feng, Zhao Jin A Real-Time Multiple Open-Circuit Fault Diagnosis Method in Voltage-Source-Inverter Fed Vector Controlled Drives. *IEEE Transactions on Power Electronics,* Feb. 2016, vol. 31, issue 2, pp. 1425–1437.

[83] Henao H., Capolino G.-A., Fernandez-Cabanas M., Filippetti F., Bruzzese C., Strangas E., Pusca R., Estima J., Riera-Guasp M., Hedayati-Kia S. Trends in fault diagnosis for electrical machines: A review of diagnostic techniques. *IEEE Ind. Electron. Mag.,* Jun. 2014, vol. 8, no. 2, pp. 31–42.

[84] Tian-Hua L., Jen-Ren F., Lipo T. A. A strategy for improving reliability of field-oriented controlledinduction motor drives. *IEEE Trans. Ind. Appl.,* Sep./Oct. 1993, vol. 29, no. 5, pp. 910–918.

[85] Diallo D., Benbouzid M. E. H., Makouf A. A fault-tolerant control architecture for induction motor drives in automotive applications. *IEEE Transactions on Vehicular Technology,* Nov. 2004, vol. 53, issue 6, pp. 1847–1855.

[86] Zagirnyak M., Melnykov V., Kalinov A. The review of methods and systems of fault-tolerant control of variable-frequency electric drives. *Przeglad Elektrotechniczny,* 2019, vol. 95, no. 1. pp. 141–144.

[87] Jannati M., Idris N. R. N., Aziz M. J. A. A new method for RFOC of Induction Motor under open-phase fault. *39th Annual Conference of the IEEE Industrial Electronics Society, IECON 2013,* 10-13 Nov. 2013, pp. 2530–2535.

[88] Jannati M., Anbaran S., Idris N. R. N., Aziz M. J. A. Vector control of three-phase induction motor under open-phase fault. *IEEE Student Conference on Research and Development (SCOReD),* 16-17 Dec. 2013, pp. 105–110.

[89] Jannati M., Monadi A., Idris N. R. N. Fault-tolerant control of 3-phase IM drive (speed-sensor fault and open-phase fault). *IEEE Conference on Energy Conversion (CENCON),* 19-20 Oct. 2015, pp. 384–389.

[90] Zagirnyak, M., A. Kalinov, V. Melnykov "Correction of the operating modes of the induction motor with damage in the stator as part of an electric drive with DTC," *Przegląd Elektrotechniczny,* Vol. 1, 12/2021, pp 151-154; https:. doi.org/10.15199/48.2021.12.30.

[91] Jannati M., Idris N. R. N., Aziz M. J. A. Speed sensorless fault-tolerant drive system of induction motor using switching extended Kalman filter. *Telkomnika Indonesian journal of electrical endineering,* 2014, vol. 12, no.11, pp.7640–7649.

[92] Odnokopylov I. G., Dementev Yu. N. Ensuring the survivability of asynchronous electric drives. *Bulletin of SUSU.* Series "Energy," 2014, vol. 14, no. 2, pp. 55–61.

[93] Odnokopylov Georgy I., Bragin A. D. Algorithms of fault tolerant control of induction motor electric drive in phase loss operate mode. *International Siberian Conference on Control and Communications (SIBCON),* 21-23 May 2015, pp. 1–5.

[94] Fu Jen-Ren, Lipo T. A. Disturbance-free operation of a multiphase current-regulated motor drive with an opened phase. *IEEE Trans. Ind. Appl.,* 1994, vol. 30, no. 5, pp. 1267–1274.

[95] Kastha Debaprasad, Bose Bimal K. Fault mode single-phase operation of a variable frequency induction motor drive and improvement of pulsating torque characteristics. *IEEE Trans. Ind. Electron,* 1994, vol. 41, no. 4, pp. 426–433.

[96] Guzmán H., Durán M. J., Barrero F. Speed control of five-phase induction motor drives with an open phase fault condition and predictive current control methods. *38th Annual Conference on IEEE Industrial Electronics Society IECON,* 25-28 Oct. 2012, pp. 3647–3652.

[97] Jannati M., Monadi A., Yen G. W., Asgari S. H., Idris N. R. N., Aziz M. J. A. A simple vector control technique for 3-phase induction motor under open-phase fault based on GA for tuning of speed PI controller. *IEEE Conference on Energy Conversion (CENCON),* 13-14 Oct. 2014, pp. 213–218.

[98] Kastha Debaprasad, Bose B. K. On-line search based pulsating torque compensation of a fault mode single-phase variable frequency induction motor drive. *IEEE Transactions on Industry Applications,* Jul/Aug 1995, vol.:31, issue 4, pp. 802–811.

[99] Odnokopylov G. I., Odnokopylov I. G., Improving the survivability of a frequency-controlled asynchronous electric drive. *Bulletin of the Tomsk Polytechnic University,* 2005, vol. 308, no. 7, pp. 143–148.

[100] Odnokopylov G. I., Bragin A. D. Vector control of a fail-safe asynchronous electric drive. *Scientific problems of transport in Siberia and the Far East,* 2013, no. 2, pp. 327–330.

[101] Barrera P. M., Leidhold R., Bossio G. R. On-Line Diagnosis of High-Resistance Connection for Inverter Fed Induction Motors. *7th IET International Conference on Power Electronics, Machines and Drives (PEMD 2014),* 8-10 April 2014, pp. 1–5.

[102] Barrera P. M., Bossio Guillermo R., Leidhold Roberto On–Line Voltage Sensorless High–Resistance Connecion Diagnosis in Induction Motor Drives. *IEEE Transactions on Industrial Electronics,* July 2015, vol.62, issue 7, pp. 4374–4384.

[103] Zarri L., Mengoni M., Gritli Y., Tani A., Filippetti F., Lee Sang Bin On-line detection of high resistance connections with inverse-sequence regulators in three phase induction motor drives. *IEEE Energy Conversion Congress and Exposition,* 15-19 Sept. 2013, pp. 4426–4431.

[104] Mengoni Michele, Zarri Luca, Gritli Yasser, Tani Angelo, Filippetti Fiorenzo, Lee Sang Bin Online Detection of High-Resistance Connections With Negative-Sequence Regulators in Three-Phase Induction Motor Drives. *IEEE Transactions on Industry Applications,* March-April 2015, vol. 51, issue 2, pp. 1579–1586.

[105] Mengoni Michele, Zarri Luca, Tani Angelo, Gritli Yasser, Serra Giovanni, Filippetti Fiorenzo, Casade Domenico On-Line Detection of High-Resistance Connections in Multiphase Induction Machines. *IEEE Transactions on Power Electronics,* Aug. 2015, vol. 30, issue 8, pp. 4505–4513.

[106] Campos-Delgado D. U., Espinoza-Trejo D. R., Palacios E. Fault-tolerant control in variable speed drives: a surveyn. *IET Electric Power Applications,* March 2008, vol. 2, issue 2, pp. 121–134.

[107] Boukhnifer Moussa, Raisemche Aziz Fault Tolerant Control for Induction Motor in Electrical Vehicle. *IEEE International Conference on Control Applications (CCA),* 3-5 Oct. 2012, pp. 136–141.

[108] Gaeid Khalaf Salloum, Ping Hew Wooi, Khalid Mustafa, Masaoud Ammar Sensor and Sensorless Fault Tolerant Control for InductionMotors Using a Wavelet Index. *Sensors,* April 2012, vol 12, issue 4, pp.4031–4050.

[109] Gaeid Khalaf Salloum, Ping Hew Wooi Wavelet fault diagnosis and tolerant of induction motor: A review. *International Journal of the Physical Sciences,* 4 February, 2011, vol. 6(3), pp. 358–376.

[110] Kouchih D., Hachelaf R., Boumalha N., Tadjine M., Boucherit M. S. Vector fault tolerant control of induction motor drives subject to stator interturn faults. *16th International Power Electronics and Motion Control Conference and Exposition (PEMC),* 21-24 Sept. 2014, pp. 108–113.

[111] Kouchih D., Hachelaf R., Boumalha N., Tadjine M., Boucherit M. S. Fault Tolerant Control of DTC Controlled Induction Motors subject to Interturn Faults. *International Conference on Control, Engineering & Information Technology (CEIT'14),* 22-23 March 2014, pp. 115–122.

[112] Melnykov, V., "The Correction of the Operation Modes for Frequency-Controlled Induction Motor with Scalar Control under Stator Windings Damage Appearance," *2021 IEEE International Conference on Modern Electrical and Energy Systems (MEES),* 2021, pp. 1-6, doi: 10.1109/MEES52427.2021.9598710.

[113] Roubache T., Chaouch S., Nait Said M. S Backstepping fault tolerant control for induction motor. *International Symposium on Power Electronics, Electrical Drives, Automation and Motion (SPEEDAM),* 18-20 June 2014, pp. 472–477.

[114] Refaat Shady S., Abu-Rub Haitham, Iqbal Atif ANN-based system for inter-turn stator winding fault tolerant DTC for induction motor drives. *17th European*

Conference on Power Electronics and Applications (EPE'15 ECCE-Europe), 8-10 Sept. 2015, pp. 1–7.

[115] Sayed-Ahmed Ahmed, Demerdash Nabeel A. O. Fault-Tolerant Operation of Delta-Connected Scalar- and Vector-Controlled AC Motor Drives. *IEEE Transactions on Power Electronics,* June 2012, vol. 27, issue 6, pp. 3041–3049.

[116] Correa M. B. R., Jacobina C. B., Mascarenhas A. W., Lima A. M. N., da Suva E. R. C. Self-compensation of unbalancing for induction motors drive system based on a quasi time-invariant model. *35th IEEE Annual Power Electronics Specialists Conference (PESC 04),* 20-25 June 2004, pp. 4390–4396.

[117] Akagi H. , Watanabe E. H., Aredes M. *Instantaneous power theory and applications to power conditioning.* – Wiley-interscience a john wiley & sons, inc. Publication, Mohamed E. El, Hawary, 2007. ISBN 978-0-470-10761-4

[118] Chumachova A., Kalinov A. Compensation of the induction motor parametric asymmetry by means of the frequency-controlled electric drive. *Tekhnichnal electrodynamika,* 2012, no. 3, pp. 87–88.

[119] Zagirnyak M., Kalinov A., Chumachova A. Correction of operating condition of a variable-frequency electric drive with a non-linear and asymmetric induction motor. *IEEE Conference transactions EUROCON 2013,* 1-4 July 2013, Croatia, pp. 1033–1037.

[120] Zagirnyak M., Kalinov A., Kostenko A. Induction motor stator windings asymmetry influence on frequency converter autonomous voltage inverter. *2nd International Conference on Intelligent Energy and Power Systems (IEPS),* June 2016, pp. 7–11.

[121] Zagirnyak M., Kalinov A., Melnykov V. Variable-frequency electric drive with a function of compensation for induction motor asymmetry. *IEEE First Ukraine Conference on Electrical and Computer Engineering (UKRCON),* 29 May-2 June 2017, pp. 338–344.

[122] Zagirnyak M., Kalinov A., Melnykov V., Kochurov I. Correction of the operating modes of an induction motor with asymmetrical stator windings at vector control. *International Conference on Electrical Drives and Power Electronics (EDPE),* 21-23 Sept. 2015, pp. 259–265.

[123] Zagirnyak M., Kalinov A., Melnykov V., Stakhiv P. Fault-tolerant control of an induction motor with broken stator electric circuit. *Electric Power Networks (EPNet),* 19-21 Sept. 2016, pp. 1–6.

[124] Melnykov V., Kalinov A. The increasing of energy characteristics of vector controlled electric drives by means of compensation for the induction motor parametrical asymmetry. *Tekhnichnal electrodynamika,* 2012, no. 3, pp. 85–86.

[125] Pronin M. V., Vorontsov A.G., Kalachikov P. N., Emelianov A. P. Electric drives and systems with electric machines and semiconductor converters (modeling, calculation, application). – St. Petersburg: Electrosila, 2004. – 252 p.

[126] Shreyner R. T. *Mathematical modeling of AC electric drives with semiconductor frequency converters.* – Yekaterinburg: URO RAS, 2000. – 654 p.

[127] Lukutin B. V. *Power Converters in Power Supply: Tutorial.* – Tomsk: TPU Publishing House, 2007. – 144 p.

[128] Zinoviev G. S. *Fundamentals of power electronics, textbook,* ed. 3rd. – Novosibirsk: ed. Of NSTU, 2004 – 672 p.

[129] Gelman M. V., Dudkin M. M., Preobrazhenskiy K. A. *Converter technology: Tutorial.* – Chelyabinsk: Publishing Center of South Ural State University, 2009. – 425 p.

[130] Semenev B. Yu. *Power electronics: professional solutions.* – Moscow: SOLON-PRESS, 2011. – 416 p.

[131] Sokolovskiy G. G. *AC drives with frequency regulation: textbook for university students.* – Moscow: Publishing Center "Academy," 2006. – 272 p.

[132] Rudakov V. V., Stoliarov I. M., Dartau V. A. *Asynchronous electric drives with vector control.* – Leningrad: Energoatomizdat, 1987. – 136 p.

[133] Pankratov V. V., Zima E. A. *Energy-optimal vector control of asynchronous electric drives: a tutorial.* – Novosibirsk: Publishing house of NSTU, 2005. – 120 p.

[134] Usoltsev A. A. *Vector Control of Induction Motors: tutorial.* – St. Petersburg: SPbGITMO (TU), 2002. – 94 p.

[135] Firago B. I., Pavliachik L. B. *Adjustable AC Drives.* – Minsk: Technoperspectiva, 2006. – 363 p.

[136] Leonard W. *Control of Electrical Drives.* – Dtrling Springer, 1996. – 420 p.

[137] Vinogradon A. B. *Vector control of AC drives.* – Ivanovo, 2008. – 298 p.

[138] Sensorless Vas P. *Vector and Direct Torque Control.* – Oxford: Oxford University Press, 1998.

[139] Popovych M. G., Lozynskyi O. Yu., Klepikov V. B. *Electromechanical control systems and electric drives: a textbook.* – Kyiv: Lubid, 2005. – 680 p.

[140] Terekhov V. M., Osipov O. I. *Electric drive control system: textbook.* – Moscow: Publishing Center "Academy," 2005. – 300 p.

[141] Peresada S. M. Vector control in an asynchronous electric drive: an analytical review. *Bulletin of DSTU,* 1999, pp. 1–23.

[142] Basharin A. V., Novikov V. A., Sokolovskiy G. G. *Control of electric drives.* – Leningrad: Energoizdat, 1982. – 392 p.

[143] Holtz J. Sensorless Control of Induction Motor Drives. *Proceedings of the IEEE,* August 2002, vol. 90, no. 8, pp. 1359–1394.

[144] Sipaylov G. A., Loos A. V. *Mathematical modeling of electrical machines (AVM): textbook for university students.* – Moscow: Vysshaia shkola, 1980. – 176 p.

[145] Chornyi O. P., Lugovoy A. V., Rodkin D. I., Susiuk G. Yu. etc. *Modeling of electromechanical systems: a textbook.* – Kremenchuk, 2001. – 410 p.

[146] Kopylov I. P. *Mathematical modeling of electrical machines.* – Moscow: Vysshaia shkola, 2001. – 327 p.

[147] Alekseev V. V., Koziaruk A. E., Zagrivnyy E. A. *Electrical machines. Modeling of electrical machines drives of mining equipment: a tutorial.* – St. Petersburg, 2006. – 58 p.

[148] Postnikov I. M. *Generalized theory and transient processes of electrical machines: a textbook for universities, second edition.* – Moscow: Vysshaia shkola, 1975. – 319 p.

[149] Shevchenko I. S., Morozov D. I. *Electromechanical processes in an asynchronous electric drive: a textbook.* – Alchevsk: Donat, 2009. – 349 p.

[150] Tolochko O. I., Razkariaka P. I., Zhurov I. O. Modeling an asynchronous motor with a stator phase failure. *Electrical and computer systems,* 2014, no. 15 (91), pp. 262–266.

[151] German-Galkin S. G. Matlab & Simulink. *Design of mechatronic systems on a PC.* – St. Petersburg: Korona-Vek, 2008. – 368 p.

[152] Kalinov A. P., Gladyr A. I. Universal teaching and research equipment for electromechanical laboratories. *Quarterly research and production journal "Electromechanical and energy saving systems,"* Kremenchuk: KSPU, 2007, vol. 1, pp. 14–19.

[153] Mykhalskyi V. M. *Means for improving the quality of electricity at the input and output of frequency and voltage converters with pulse-width modulation.* – Kyiv, Institute of Electrodynamics of the National Academy of Sciences of Ukraine, 2013. – 340 p.

[154] Bagarty D. P., Dash S. S., Bastia S., Sahu P. Modified Sine Triangle Approach For 30-Degree Split Clamp Pulse Width Modulation For Three Phase Voltage Source Inverter. *First International Conference on Power, Control and Computing Technologies (ICPC2T),* 2020, pp. 171–175. doi: 10.1109/ICPC2T48082.2020.9071467.

[155] Peresada S., Kovbasa S., Toniellli A. Theoretical comparison of indirect field-oriented control of induction motors. *Bulletin of Kremenchug State Polytechnic University,* 2002, vol. 1, pp. 43–49.

[156] Potapenko E. M., Potapenko E. E. *Robust algorithms for vector control of an asynchronous drive.* – Zaporozhye: ZNTU, 2009. – 352 p.

[157] Melnykov V. O., Kalinov A. P., Ogar V. O. Research of indicators of quality of energy conversion in the electric drive with vector control. *Bulletin of NTU "KhPI," "Problems of automated electric drive. Theory and practice,"* Kharkiv: NTU "KhPI," 2010, no. 28, pp. 230–235.

[158] Kalinov A. P., Mamchur D. G., Chumacheva A. V. Energy characteristics of asynchronous motors in dynamic operating modes. *Collection of scientific works of Dniprodzerzhynsk State Technical University: (technical sciences),* Dniprodzerzhynsk: DSTU, 2009, no. 1 (11), pp. 104–110.

[159] Aliev Ya. A. Influence on the operation of a synchronous motor of higher harmonics when it is supplied with a non-sinusoidal voltage. *Electromagnetic processes in drives with frequency control, 1972,* Leningrad: Nauka, 1972, pp. 41–53.

[160] Volkov A. V., Skalko Yu. S. Mathematical model of total power losses in a frequency-controlled asynchronous electric drive. *Electrical engineering and electric power, Zaporizhzhia,* 2006, no. 2, pp. 22–27.

[161] Pivniak G. G., Volkov A. V. *Modern frequency-controlled asynchronous electric drives with pulse-width modulation.* – Dnepropetrovsk: NSU, 2006. – 470 p.

[162] Voronin P. A. *Power semiconductor switches: families, characteristics, applications,* second edit., revised and additional. – Moscow: Dodeka-XXI Publishing House, 2005. – 384 p.

[163] Meleshin V. I. *Transistor converter technology.* – Moscow: Tekhnosfera, 2006. – 632 p.

[164] Semenev B. Yu. *Power electronics: from simple to complex.* – Moscow: Solon-Press, 2008. – 416 p.

[165] Melnykov V., Perdul'ak J. Experimental investigations of working modes of power thyristor keys in semiconductor converters. *13th Scientific Conference of Young Researchers of Faculty of Electrical Engineering and Informatics Technical University of Košice,* 2013. pp. 18–321.

[166] Melnykov V. Energy loss in power keys of frequency-controlled electric drive under unsymmetry of induction motor windings. *Proceeding of scientific and student's works in the field of Industrial Electrical Engineering, SSIEE,* 2015, vol. 4, pp. 185–188.

[167] Abrahamsen F., David A. Adjustable speed drive with active filtering capability for harmonic current compensation. *Conf. Rec. of IEEE–PESC,* 1995, vol. 2. – pp. 1137–1143.

[168] Zyl A., Enslin J. H. R., Spee R. Converter-based solution to power quality problems on radial distribution lines. *IEEE Transactions of Industrial Applicaations,* 1996, vol. 32, no 6, pp. 1323–1330.

[169] Sinchuk O. N., Polishchuk P. I., Chornyi A. P., Pasko O. V. Research of processes and development of an energy conversion control algorithm in an asynchronous traction electric drive with frequency conversion of supply voltage regulation. *Bulletin of KSPU. Transactions of KSPU,* 2006, vol. 3 (38), part. 2, pp. 17–20.

[170] Sychev S. D., Chornyi A. P. Control of the process of energy conversion in an induction motor. *Collection of works of the international scientific and technical conference "Electromechanical systems, methods of modeling and optimization,"* Kremenchuk: KDPU, 2003, pp. 95–95.

[171] Kalinov A., Chumachova A. Approach to the compensation of the induction motor parameters asymmetry by means of the frequency–controlled electric drive. *XIII International PhD Workshop OWD'2011. Conference Arhives PTETiS,* Poland, 2011, no. 29, pp. 402–406.

[172] Kalinov A. P., Chumachova A. V. Substantiation of the approach to compensation of parametric asymmetry of the induction motor as a part of the frequency-regulated electric drive. *Electrical engineering and electric power,* 2012, no. 1, pp. 40–44.

[173] Kravchik A. E., Shlaf M. M., Afonin V. I., Sobolenskaya E. A. *4A Series Induction Motors: Handbook.* – Moscow: Energoizdat, 1982. – 504 p.

[174] Chumachova A. V., Kalinov A. P. Correction of the mode of operation of the frequency-regulated electric drive with the asymmetric asynchronous motor. *"Electromechanical and energy saving systems,"* 2012, no. 4(20), pp C. 30–35.

[175] Baykov A. I. *Mathematical modeling of elements and systems of an automated electric drive: a tutorial.* – Nizhny Novgorod: Nizhny Novgorod R. E. Alekseev State Technical University, 2008. – 173 p.

[176] Melnykov V. O., Kalinov A. P., Mamchur D. G., Ogar V. O. Experimental estimation of validity the mathematical model of induction motor with stator

windings asymmetry. *Electromechanical and energy saving systems,* 2012, vol. 2(18), pp. 31–35.

[177] Jovanovic M., Crow D. E. Merits and limitations of full–bridge rectifier with LC filter in meeting IEC 1000–3–2 harmonic–limit specifications. *Conf. Rec. of IEEE–APEC,* 1996, pp. 354– 360.

[178] Zhemerov G. G., Sokol E. I., Krilov D. S., Baru A. Yu., Shindnes Yu. L. Computer simulation of a converter system with a compensated controlled rectifier. *Technical electrodynamics,* 2002, no.4, pp. 59–62.

[179] German-Galkin S. G. Matlab & Simulink. *Design of mechatronic systems on a PC.* – St. Petersburg: Korona-Vek, 2008. – 368 p.

[180] Kalinov A. P., Mamchur D. G. Mathematical models for research the effect of structural asymmetries of electrical machines on their electromagnetic parameters. *Transactions of Kremenchug Mykhailo Ostrohradskyi State Polytechnic University,* 2007, vol. 3 (44), part. 2, pp. 150–154.

[181] Chumachova A. V., Kalinov A. P. Compensation of the induction motor parametric asymmetry by means of the frequency-regulated electric drive. *Technical electrodynamics,* 2012, no. 3, pp. 87–88.

[182] Stone G. C., Sedding H. G., Lloyd B. A., Gupta B. K. The ability of diagnostic tests to estimate the remaining life time of stator winding insulation. *IEEE Transactions on Energy Conversion,* 1998, vol. 3, no. 4, pp. 833–841.

[183] Zherve G. K. Industrial testing of electrical machines. – Leningrad: Energoatomizdat, 1984. – 408 p.

[184] Kotelenets N. F., Kuznetsov N. L. *Testing and reliability of electrical machines.* – Mockow: Vysshaia shkola, 1988. – 231 p.

[185] Kostenko A. V., Kalinov A. P., Kolosiuk V. P., Gurzhii A. M. Research of an energy loss in power keys of frequency-controlled electric drive with scalar control under the compensation of an induction motor unsymmetry. *Transactions of Kremenchuk Mykhailo Ostrohradskyi National university,* Kremenchuk, 2015, no. 3 (63), part. 1, pp. 48–53.

[186] Domnin I. F., Zhemerov G. G., Ilina O. V., Tugay D. V. Instantaneous active power ripple compensation in circuits with resistive load. *Technical electrodynamics,* 2006, no. 6, pp. 36–41.

[187] Domnin I. F., Zhemerov G. G., Krulov D. S., Sokol E. I. Modern power theories and their use in power electronics converter systems. *Technical electrodynamics,* 2004, no. 4, pp. 80–91.

[188] Zhezhelenko I. V., Saenko Yu. L. *Issues of power quality in electrical installations.* – Mariupol: PSTU, 1996. – 173 p.

[189] Barabashchuk V. I., Kredentser B. P., Miroshnichenko V.I. *Experiment planning in engineering* [edited by Kredentser B. P.] – Kyiv: Technics, 1984. – 200 p.

[190] Gemke R. G. *Malfunctions of electrical machines* [edited by Umantsev R.B.] – Leningrad: Energoatomizdat, 1989. – 336 p.

[191] Anistratenko V. O., Fedorov V. G. *Mathematical planning of experiments in agro-industrial complex: tutorial.* – Kyiv: Vysshaia shkola, 1993. – 375 p.

[192] Muhachev V. A. *Planning and Processing Experiment Results: tutorial.* – Tomsk: Tomsk State University of Control Systems and Radioelectronics, 2007. – 118 p.

[193] Chumacheva A. V. Adaptive compensation system of induction motor unsymmetry by means of variable-frequency electric drive. *Electromechanical and energy saving systems,* Kremenchuk: KrNU, 2013, vol. 2(22), part 2, pp. 128–134.

[194] Tereshkevich L. B., Chervinskaya T. N. Determination of the limit values of the parameters of the asymmetry of the electric mode from the condition of permissible heating of an asynchronous motor. *Scientific works of VNTU,* 2011, vol. 3, pp. 43–48.

[195] Kliuchev V. I. *Electric drive theory.* – Moscow: Energoatomizdat, 1988. – 704 pc.

[196] Syromyatnikov I. A. *Operating modes of asynchronous and synchronous motors.* – Moscow: Energoatomizdat, 1984. – 240 p.

[197] Chornyi A. P., Lugovoy A. V., Maksimov M. N., Rodkin D. I., Sisyuk G. Yu. Operational reliability of AC electric motors and ways to improve it. *Problems of creating new machines and technologies: Scientific works of KSPU,* 2000, vol.1 (8), pp.150–156.

[198] Bessonov L. A. *Theoretical foundations of electrical engineering. Electrical Circuits: Textbook, 10th ed.* – Moscow: Gardariki, 2001. – 638 p.

[199] Demirchyan K. S., Neyman L. R., Korovkin N. V. T*heoretical foundations of electrical engineering: a textbook for universities, 5th ed.* – St. Petersburg: Piter, 2009. – 512 p.

[200] Melnykov V. O., Kalinov A. P., Kochurov I. M. Compensation for asynchronous stator winding asymmetry in vector control systems. *Electrical engineering and electric power, Zaporozhye: ZNTU,* 2015, no. 1, pp. 26–33.

[201] Soares V., Verdelho P., Marques G. D. An instantaneous active and reactive current component method for active filters. *IEEE Transactions on Power Electronics,* July 2000, vol. 15, no. 4, pp. 660–669. doi: 10.1109/63.849036.

[202] Akagi H., Kanazawa Y., Nabae A. Instataneous reactive power compensators comprising switching devices without energy storage components. *IEEE Trans. Ind. Applicat.,* May/June, 1984, vol. 20, pp. 625–630.

[203] Bitoleanu Alexandru, Popescu Mihaela, Suru Vlad P–q power theory: Some theoretical and practical aspects. *2010 International School on Nonsinusoidal Currents and Compensation,* 15-18 June 2010, pp. 74–79.

[204] Zagirnyak M. V., Kalinov A. P., Maliakova M. S. An algorithm for electric circuits calculation based on instantaneous power component balance. *Przegląd elektrotechniczny (Electrical Review),* 2011, R. 87 NR 12b/2011, pp. 212–215.

[205] Kozyaruk A. E., Rudakov V. V. *Modern and perspective algorithmic support of frequency-controlled electric drives.* – Saint Petersburg: St. Petersburg Electrotechnical Company, 2004. – 127 p.

[206] Kubota H., Matsuse K. Regenerationg-Mode Low-Speed Operation of Sensorless Induction Motor Drive With Adaptive Observer. *IEEE Transactions on Industry Applications,* July/August 2002, vol. 38, no. 4, pp. 1081–1086.

[207] Melnykov V. O. Using phase control systems for correction of asymmetric modes of operation of asynchronous electric drive with vector control. Electromechanical and energy saving systems. *Quarterly research and production journal,* Kremenchuk: KrNU, 2015, vol. 2/2015 (30), pp. 45–51.

[208] Maltseva O. P., Udut L. S., Koyanin N. V. *Control systems for asynchronous frequency-controlled electric drives: a tutorial.* – Tomsk: Publishing House of Tomsk Polytechnic University, 2011. – 476 p.

[209] Dyadik V. F., Baydali S. A., Krinitsyn N. S. *Theory of automatic control: textbook.* – Tomsk: Publishing House of Tomsk Polytechnic University, 2011. – 196 p.

[210] Zagirnyak M., Kalinov A., Melnykov V. Sensorless vector control systems with the correction of stator windings asymmetry in induction motor. *Przegląd elektrotechniczny,* 2013, R. 89 NR 12/2013, pp. 340–343.

[211] Zagirnyak M., Kalinov A., Melnykov V. Decrease of the thermal overloads of a variable-frequency electric drive at damages in the electric circuit of an induction motor stator. *Przegląd elektrotechniczny,* 2019, no. 95 (5), pp. 43–46.

[212] Pankratov V. V., Zima E. A., Maslov M. O. On the problem of building "intelligent" frequency converters for general industrial AC drives. *Proceedings of the VII International Conf. "Actual problems of electronic instrumentation" APEP-2004* (Novosibirsk, September 21-24, 2004), Novosibirsk: NSTU, vol. 6, pp. 143–145.

[213] Vinogradov A. B., Sibirtsev A. N., Kolodin I. Yu. Adaptive-vector control system for sensorless asynchronous electric drive EPV series. *Power electronics,* 2006, no. 3, pp. 50–55.

[214] Pankratov V. V., Maslov M. O. Problems of synthesis of identification algorithms for sensorless asynchronous electric drives with vector control and a variant of their solution. *Power intelligent electronics,* 2007, no. 1(6), pp. 10–22.

[215] Gastli A., Matsui N. Stator flux controlled V/f PWM inverter with identification of IM parameters (induction motors). *IEEE Transactions on Industrial Electronics,* Aug. 1992, vol. 39, no. 4, pp. 334-340. doi: 10.1109/41.149751.

[216] Gastli A. Identification of induction motor equivalent circuit parameters using the single-phase test. *IEEE Transactions on Energy Conversion,* March 1999, vol. 14, no. 1, pp. 51–56. doi: 10.1109/60.749147.

[217] Vinogradov A. B., Chistoserdov V. L., Sibirtsev A. N., Monov D. A. Asynchronous electric drive for general industrial use with direct digital control and advanced intellectual properties. *Izvestiya vuzov. Electromechanics,* 2001, no. 3, pp. 60–67.

[218] Vinogradov A. B., Monov D. A., Kolodin I. Yu. Automation of parameter setting in an asynchronous electric drive with microprocessor control. *The thesis of the report of the international scientific and technical conference "Status and prospects for the development of electrical technology,"* Ivanovo, 1999, p. 197.

[219] Braslavsky I. Ya., Ishmatov Z. Sh., Barats E. I. Adaptive Direct Torque Control System for Induction Motor. *Electrical engineering,* 2001, no. 11, pp. 36–40.

[220] Chepkunov R. A., Kalonov A. P., Melnykov V. O. Determining the parameters of an asynchronous motor when working with a frequency converter in auto-tuning mode. *Scientific practices of DonNTU. Series "Electrical engineering and energy,"* Donetsk: DonNTU, 2012, issue 1 (12)-2 (13), pp. 250–254.

[221] Vinogradov A. B., Sibirtsev A. N., Zhuravlev S. V. Sensorless electric drive of hoisting and transport mechanisms. *Power electronics,* 2007, no. 1, pp. 46–52.

[222] Vinogradov A. B., Chistoserdov V. L., Sibirtsev A. N. Adaptive vector control system for asynchronous electric drive. *Electrical engineering,* 2003, no. 7, pp. 7–17.

[223] Maslov M. O., Pankratov V. V. One Algorithm for Pre-Identification of Parameters for an Induction Drive with Vector Control. *Proceedings of the XIII International Conference "Alternating Current Electric Drives"* (EPPT'05, March 15–18, 2005, Yekaterinburg, Russia), Yekaterinburg, 2005, pp. 99–102.

[224] Beshta A. S., Semin A. A. Peculiarities of using a t-shaped equivalent circuit of an asynchronous motor for the tasks of parameter identification. *Electromechanical and energy saving systems. Thematic issue "Problems of automated electric drive. Theory and Practice,"* Kremenchuk, 2012, issue 3/2012 (19), pp. 553–555.

[225] Kalinov A. P., Rieznik D. V., Rodkin D. I., Romashikhin Iu. V. Determination of electromagnetic parameters of asynchronous motors when powered by low-frequency voltage. *Bulletin of the KSPU, Kremenchuk, KSPU,* 2006, issue 4/2006 (39), part 1, pp. 117–121.

[226] Rieznik D. V., Rodkin D. I., Romashikhin Iu. V. Features of determining the electromagnetic parameters of asynchronous motors when using low-frequency test voltage. *AC drives: Proceedings of the international 14 NTC, Yekaterinburg: GOU VPO USTU-UPI,* 2007, pp. 279–284.

[227] Melnykov V. O., Kalinov A. P. Method for induction motors parameters identification with low-frequency feed of stator windings. *10th Scientific Conference of Young Researchers of Faculty of Electrical Engineering and Informatics Technical University of Košice,* 2010, pp. 379–381.

[228] Rieznik D. V., Rodkin D. I., Chornyi A. P. To the determination of the parameters of asynchronous motors at different frequency test voltage. *Electromechanical and energy saving systems. Thematic issue "Problems of automated electric drive. Theory and Practice,"* Kremenchuk, 2012, issue 3/2012 (19), pp. 556–561.

[229] Kalitkin N. N. *Numerical Methods.* – Moscow: Nauka, 1978. – 168 p.

[230] Bronshteyn I. N., Semendyaev K. A. *Handbook of Mathematics for Engineers and University Students.* – Moscow: Nauka, 1981. – 720 p.

[231] Porshnev S. V., Belenkova I. V. *Numerical methods based on Mathcad.* – Saint Petersburg: BHV-Petersburg, 2005. – 464 p.

[232] Kalinov A. P. Melnykov V. O. Substantiation of initial approximations in the identification of electromagnetic parameters of an induction motor. *Electromechanical and energy saving systems,* Kremenchuk: KSPU, 2009, issue 1/2009(5), pp. 57–59.

[233] Zelenov A. B. *Electric Drive Theory. Part II: tutorial.* – Alchevsk: DonSTU, 2005. – 513 p.

[234] Volkov E. A. *Numerical Methods: textbook for universities* [2 edit., correct.]. – Moscow: Nauka,. Ch. ed. Phys.-Math. litas., 1987. – 248 p.

[235] Rakitin V. I., Pervushin V. B. *A practical guide to computing methods with the application of programs for personal computers: tutorial.* – Moscow: Vysshaia shkola, I998. – 383 p.

[236] *Rotating electrical machines. General test methods:* GOST 11828-86. – Moscow: IPK Standards Publishing House, 2003. – 32 p.

[237] Rodkin D. I., Zdor I. E. Features of determining the parameters of the equivalent circuit of an asynchronous motor when powered from a three-wire line. *Bulletin of the KSPU, Kremenchuk: KSPU,* 2001, issue 1, pp. 212–216.

[238] Rodkin D. I., Kalinov A. P., Romashikhin Iu. V. Estimation of the effectiveness of diagnostic methods for AC motors. *Bulletin of the KSPU, Kremenchuk: KSPU,* 2006, issue 3 (38), part 2, pp. 77–82.

[239] Chornyi A. P., Rodkin D. I., Sidorenko V. N., Kalinov A. P. Determination of IM parameters from a continuous function of the stator current with discrete time. *Bulletin of the KSPU, Kremenchuk: KSPU,* 2004, issue 3 (26), pp. 123–127.

[240] Goroshkov B. I. *Radioelectronic devices: reference book.* – Moscow: Radio and communication, 1984. – 400 p.

[241] Johnson D., Johnson J., Moore G. *Guide to active filters: translation from English.* – Moscow: Energoatomizdat, 1983. – 128 p.

[242] Sidorenko V. N., Chornyi A. P., Kalinov A. P. Improving the accuracy of determining the electromagnetic parameters of induction motors using signal correction algorithms. *Bulletin of the KSPU, Kremenchuk: KSPU,* 2006, issue 3/2006 (38), part 2, pp. 88–92.

[243] Abramov V. M. *Electronic elements of automatic control devices: diagrams, calculation, reference data.* – Moscow: ICC "Akademkniga," 2006. – 680 p.

[244] Gutnikov V. S. *Filtering of measuring signals.* – Leningrad: Energoatomizdat. Leningrad. department, 1990. – 192 p.

[245] Kalinov A. P., Melnykov V. O. Identification of electromagnetic parameters of an induction motor when using low-frequency power supply. *Technical electrodynamics. Thematic issue "Problems of modern electrical engineering,"* Kyiv: Institute of Electrodynamics of the National Academy of Sciences of Ukraine, 2010, part 2, pp. 138–143.

[246] Kalinov A. P., Vorobeychik O. S., Melnykov V. A. Measurement of currents and voltages of an asynchronous motor as part of a frequency-controlled electric drive. *Collection of scientific works of Dniprodzerzhynsk State Technical University: (technical sciences),* Dniprodzerzhynsk: DSTU, 2009, issue № 1 (11), pp. 110–117.

[247] Kovacova, I. and V. Melnykov, "Magnetic Fields of DC PM Motor and its EMC," 2019 *IEEE International Conference on Modern Electrical and Energy Systems (MEES),* 2019, pp. 186-189, doi: 10.1109/MEES.2019.8896558.

[248] Melnykov V. O., Kalinov A. P. Investigation of errors in identification of electromagnetic parameters of induction motors at variable power frequency. *Bulletin of Kremenchuk Mykhailo Ostrohradskyi National University,* Kremenchuk: KNU, 2010, issue 4/2010 (63), part 1, pp. 125–128.

[249] Rieznik D. V., Rodkin D. I., Kalinov A. P. Improving the accuracy of determining the electromagnetic parameters of asynchronous motors using low-frequency test voltage. *Bulletin of the KSPU, Kremenchuk: KSPU,* 2007, issue 3 (44), part 2, pp. 144–149.

[250] Sidorenko V. N., Chornyi A. P. Features of measuring and analyzing signals in DC drives. *Bulletin of the KSPU, Kremenchuk: KSPU,* 2001, issue 1, pp. 340 –344.

[251] Prus V. V., Rodkin D. I. Research of the electrical and magnetic properties of electrical steel after thermal effects. *Scientific works of the Kremenchug State Polytechnic University "Problems of creating new machines and technologies,"* issue 2/2000 (9), pp. 228–234.

[252] Chumachova A. V., Kalinov A. P. Determination of the operation optimal period of an induction motor by the minimum costs criterion. *Bulletin of Kremenchuk Mykhailo Ostrohradskyi National University,* Kremenchuk: KNU, 2010, issue 4 (63), part 2, pp. 142–146.

About the Authors

Mykhaylo Zagirnyak
Rector, DSc (Eng.), Professor
Kremenchuk Mykhailo Ostrohradskyi National University,
Kremenchuk, Ukraine
Email: mzagirn@kdu.edu.ua;mzagirn@gmail.com

Mykhaylo V. Zagirnyak is a Professor and Rector of Kremenchuk Mykhailo Ostrohradskyi National University (Ukraine), a Full Member of the National Academy of Pedagogical Sciences of Ukraine, and an IEEE Senior Member. He is also a member of many editorial boards and scientific boards at numerous international conferences. His research interests include the calculation of magnetic fields for electromagnetic separators, electric machines, and other electromagnetic devices. He is the author and coauthor of more than 700 research papers, including thirteen monographs and sixteen textbooks.

Andrii Kalinov
Associate Professor, PhD (Eng.).
Kremenchuk Mykhailo Ostrohradskyi National University,
Kremenchuk, Ukraine
Email: andrii.kalinov@gmail.com

Andrii P. Kalinov is an Associate Professor for the Department of the Systems of Automatic Control and Electric Drive at Kremenchuk Mykhailo Ostrohradskyi National University (Ukraine). He also is the Technical Chief of TOV NVP ENERGO-PLYUS. His research interests include electric drives, induction motors diagnostic and electrical mechanism. He is the author and coauthor of more than 260 research papers, including three monographs and three textbooks.

Anna Kostenko
Assistant, PhD (Eng.)
Kremenchuk Mykhailo Ostrohradskyi National University,
Kremenchuk, Ukraine
Email: annakostenko.knu@gmail.com

Anna V. Kostenko is an Assistant for the Department of the Systems of Automatic Control and Electric Drive at Kremenchuk Mykhailo Ostrohradskyi National University (Ukraine). Her research interests include assessment of energy efficiency of induction motors of industrial electric drive systems. She is the author and a coauthor of more than 30 research papers, including one monograph.

Viacheslav Melnykov
Associate Professor, PhD (Eng.)
Kremenchuk Mykhailo Ostrohradskyi National University,
Kremenchuk, Ukraine
Email: melnykow@gmail.com

Viacheslav O. Melnykov is an Associate Professor for the Department of the Systems of Automatic Control and Electric Drive at Kremenchuk Mykhailo Ostrohradskyi National University (Ukraine). His research interests include development of methods and systems of fault-tolerant control of frequency-controlled electric drives. He is the author and coauthor of more than 55 research papers, including two textbooks.

Index

P

R

S

V